SEMICONDUCTOR DEVICES FOR POWER CONDITIONING

Earlier Brown Boveri Symposia

Flow Research on Blading • 1969
Edited by L. S. Dzung

Real-Time Control of Electric Power Systems • 1971
Edited by E. Handschin

High-Temperature Materials in Gas Turbines • 1973
Edited by P. R. Sahm and M. O. Speidel

Nonemissive Electrooptic Displays • 1975
Edited by A. R. Kmetz and F. K. von Willisen

Current Interruption in High-Voltage Networks • 1978
Edited by Klaus Ragaller

Surges in High-Voltage Networks • 1980
Edited by Klaus Ragaller

SEMICONDUCTOR DEVICES FOR POWER CONDITIONING

Edited by

Roland Sittig

BBC Brown, Boveri & Company, Limited
Birr, Switzerland

and

P. Roggwiller

BBC Brown, Boveri & Company, Limited
Baden, Switzerland

PLENUM PRESS • NEW YORK AND LONDON

Library of Congress Cataloging in Publication Data

Brown Boveri Symposium on Semiconductor Devices for Power Conditioning
(1981: Brown Boveri Research Center)
  Semiconductor devices for power conditioning.

Bibliography: p.
  Includes index.
  1. Power semiconductors—Congresses. 2. Semiconductor rectifiers—Congresses.
I. Sittig, Roland. II. Roggwiller, P. III. Title.
TK7881.15.B76 1981   621.31'7   82-16167
ISBN 0-306-41131-8

Proceedings of the Brown Boveri Symposium on Semiconductor Devices
for Power Conditioning, held at the Brown Boveri Research Center
September 21–22, 1981, in Baden, Switzerland

©1982 Plenum Press, New York
A Division of Plenum Publishing Corporation
233 Spring Street, New York, N.Y. 10013

Printed in the United States of America

FOREWORD

The Brown Boveri Symposia are by now part of firmly established tradition. This is the seventh event in a series which was initiated shortly afer Corporate Research was established as a separate entity within our Company; the Symposia are held every other year. The themes to date have been

1969	Flow Research on Blading
1971	Real-Time Control of Electric Power Systems
1973	High-Temperature Materials in Gas Turbines
1975	Nonemissive Electrooptic Displays
1977	Current Interruption in High-Voltage Networks
1979	Surges in High-Voltage Networks
1981	Semiconductor Devices for Power Conditioning

Why have we chosen these titles? At the outset we established certain selection criteria; we felt that a subject for a Symposium should fulfill the following requirements:

- It should characterize a part of a thoroughly scientific discipline; in other words, it should describe an area of scholarly study and research.

- It should be of current interest in the sense that important results have recently been obtained and considerable research effort is underway in the international scientific community.

- It should bear some relation to the scientific and technological acitivity of our Company.

Let us look at the requirement "current interest": Some of the topics on the list have been the subject of research for several decades, some even from the beginning of the century. One might wonder, then, why such fields could be regarded as particularly timely in the 1960s and 1970s. A few remarks on this subject therefore are in order.

It does no harm to be reminded occasionally that modern technology still contains large elements of empiricism. It can be stated almost as a rule that the older the art the greater the empirical content. The oldest manufacturing art without any doubt is ceramics, dating back at least 6,000 years; and today, despite the enormous advances in solid-state physics and in the material sciences, the field of ceramics still contains much empirical knowledge which is not based on scientific understanding. At the other end of the spectrum is microelectronics: This represents an achievement of the last 30 years; the phenomena underlying electronic devices have been predicted on the basis of scientific analysis or have, at least, been well understood scientifically.

Most subjects of our Symposia fall between these two extremes. They concern fields that are old, but not ancient. Experience shows that scientific progress in such areas does not occur at a constant rate - it comes in waves that are often followed by a standstill. The waves are often sparked by an external event that may come from quite unexpected sources. For example, the discovery of a new mathematical method may open the possibility for the numerical solution of a class of problems that hitherto were intractable; or drastically increased energy prices may trigger a renewed effort towards the improvement of the efficiency of energy converters. Our Symposium subjects have been chosen so as to coincide with such a wave.

This year's subject is a good illustration for such a sequence of events: Electronic power conditioning was taken up by Brown Boveri almost 70 years ago. Of course, neither the word "electronic" nor the term "power conditioning" was used at that time; they were coined several decades later. In 1914, a mercury arc rectifier for 165 V and 100 A was developed and manufactured; it was used for the charging of storage batteries. By today's definition the mercury arc rectifier is an electronic device. It was the object of rather intensive research activity at Brown Boveri. An important breakthrough was the introduction of the control grid in 1926. The grid made it possible not only to supply a control voltage, by varying the moment of firing, but also to achieve d.c. to a.c. conversion. Further development led to the first high voltage d.c. transmission from Baden to Zurich in 1939. This experimental installation, operating at 50 kV, was a pioneering achievement. The subsequent world war brought development to a halt for a number of years. The installation anticipated the course of technology by a quarter of a century.

Shortly after the war, a strange thing happened: Engineers discovered that a.c. could be converted to d.c. simply by opening and closing mechanical contacts at the right instant. Simple as this idea sounds, it had apparently never even been tested, yet alone carried to the point of commercial feasibility. The contact rectifier was used in those applications where low voltages and very high currents were used; such is the case, for example, in electrolysis. The low resistance of the mechanical contacts provided an efficiency far superior to the mercury arc. With the contact rectifier, we experienced a reversal of the prevailing trend: electronics was replaced by electromechanics.

This reversal lasted for about 15 years, i.e. from 1945 to 1960. Then, with the advent of power semiconductors, the era of contact rectifiers as well as mercury arc devices abruptly came to an end. An area of rapid progress, based on solid state physics, began and is still in full swing.

It was one of the purposes of this Symposium to assess where we stand today with regard to power semiconductor technology. The high level of activity in this field is reflected in the most gratifying response to the announcement of our Symposium. As in previous meetings, we limited the number of participants to about 100 in order to maintain the character of a specialist meeting of restricted size, and it was with much regret that we were forced to disappoint many who wished to participate but who could not be accommodated. We hope that the publication of the present volume is a partial consolation for those whom we did not have the pleasure of welcoming as our guests.

The Symposium was attended by 97 participants from 12 countries. It was both an honor and a pleasure to welcome these scientists and engineers from so many different parts of the world. Their willingness to travel to Baden and spend two full days with us was a challenge as well as an obligation to us as organizers, and we sincerely hope that the expectations which prompted them to attend were fulfilled.

To conclude, we should like to take this opportunity to express our sincere gratitude to every participant in the Symposium. We hope they consider the time spent with us to have been worthwile. Thanks are due primarily to the authors for having spared no effort in preparing their papers: the contents of this volume reflect the high quality of their work. We thank also the participants in the discussions, both formal and inform- al, and the editors of these proceedings.

The selection of the theme, the layout of the program and contact with the speakers were the responsibility of Dr. Roland Sittig, development manager in the Semiconductor Components Division of Brown Boveri. His careful and competent preparation in col- laboration with Dr. R. P. Roggwiller and Dr. C. Schüler, head of the Solid State Sciences Department at the Research Center, has been instrumental in the success of the Symposium. Our thanks should also go to Mr. H. Wilhelm and his staff for the smooth running of the administrative side of the meeting.

A. P. Speiser
Director of Research

PREFACE

Semiconductor devices have strongly influenced the development of installations for power conditioning over the last two decades. As a result of an ever increasing rate of improvement of the characteristics of these devices, they have been able to successfully compete with rotating machines and mercury-vapor convertors.

In recent years two factors have provided additional impetus for the development of semiconductor power devices: the ability to numerically simulate the device behavior during actual operating conditions and the ability to produce device structures incorporating nearly all features which are thought beneficial for their electrical characterstics.

These two aspects have permitted a more rapid rate of improvement of the more conventional power devices such as thyristors and bipolar power transistors, and to realization of new and more sophisticated devices such as gate turn-off thyristors, junction field effect devices, and power MOSFETs.

The competition between different devices for the same application led to a situation where further progress in device development and circuit design became strongly interdependent and therefore required that these two aspects no longer be considered independently.

The goals of this Symposium were to present the state of the art of semiconductor power device technology and to assess this from the aspect of circuit needs. In the tradition of the Brown Boveri Symposia, these goals were achieved through survey lectures given by leading scientists and engineers in the field.

The Symposium, which took place on September 21 and 22, 1981, at the Brown Boveri Research Center in Baden, Switzerland, comprised three main areas of interest:

The area of very high power at line frequency is still governed by very large thyristors, which can be supplemented by either external or integrated light-fired auxiliary thyristors to permit potential-free triggering.

Two different alternatives were considered in the area of high power at higher frequencies: devices exhibiting an extremely short recovery time such as reverse-conducting thyristors and devices offering switch-off capabilities such as gate turn-off thyristors.

At the present time the most dramatic development is taking place in the third area, namely, that of lower voltage and high frequencies. Four lectures were devoted to the progress made in the development of fast devices.

A discussion period followed each individual paper and at the end of each of the three main areas of interest. An edited version of the discussions has been included in this volume.

At this point we would like to express our sincere thanks to everyone who contributed to the success of the Symposium. We are especially greateful to the authors of the papers, each of whom promptly accepted our invitation to participate and maintained a strong interest from that starting point up to the preparation of this volume. The expertise of the four session chairmen contributed a great deal to the success of the discussion periods and is greatefully acknowledged.

It is a pleasure to thank Dr. A.M. Escudier who relieved us of the linguistic aspects of the editorial work and also assisted us in the proofreading. Miss D. Burkhalter and Miss U. Wirth were responsible for the extensive secretarial work associated with the Symposium. The entire camera-ready manuscript for this volume was prepared by Miss U. Wirth, and several figures were redrawn by Mrs. M. Zamfirescu. The help of all these individuals is gratefully acknowledged.

Baden, May 1982

R. Sittig   P. Roggwiller

CONTENTS

PARTICIPANTS

Dipl.-Ing. C. Abbas
BBC Brown,Boveri & Co.
Research Center
CH-5405 Baden
Switzerland

Dr. L. Abraham
Brown, Boveri & Cie. AG
Dept. ZEK/E
Neustadter Strasse 62
D-Mannheim-Käfertal (Süd)
West Germany

Dr. Michael S. Adler
General Electric
Corporate R & D
Building K-1, Room 1521
Schenectady, N.Y. 12345
U.S.A.

Mr. J.C. Allacoque
CEM-Division Entrainements et
Sous-Ensembles Electroniques
FEL
Av. du Bel Air
F-69627 Villeurbanne Cedex
France

Dr. F.D. Althoff
Brown, Boveri & Cie. AG
VST
Postfach 3 51
D-6800 Mannheim 1
West Germany

Dr. Peter Appun
Brown, Boveri & Cie. AG
Dept. VK
Neustadter Strasse 62
D-6800 Mannheim 31
West Germany

Mr. Martin Argast
Blocher-Motor GmbH & Co
Dieselstr. 4
D-7430 Metzingen
West Germany

Mr. Carl Banic
United Technologies Power Systems
P.O. Box 109 MS 12
South Windsor, CT 06074
USA

Dr. Hans W. Becke
Bell Laboratories
600 Mountain Ave.
Murray Hill, N.J. 07974
U.S.A.

Dr. Günther H. Berndes
Brown, Boveri & Cie. AG
Dept. HS/LE
Boveristrasse 1
D-6840 Lampertheim
West Germany

Dr. H. Birnbreier
Brown, Boveri & Cie. AG
Zentrales Forschungslabor
Eppelheimer Str. 82
D-6900 Heidelberg
West Germany

Mr. Henri Bonhomme
Service du Prof. Legros
Université de Liège
Service d'Electrotechnique
Rue St-Gilles 33
B-4000 Liège
Belgique

Mr. Hansjürg Bossi
BBC Brown, Boveri & Co.
Dept. IE
CH-5401 Baden
Switzerland

Mr. Kurt Brisby
AESEA AB
S-721 83 Västeras
Sweden

Ing. Claude Caën
CEM-Cie Electro-Mécanique
(CERCEM)
49, rue du Commandant Rolland
F-93350 Le Bourget
France

Dr. Nigel Coulthard
Thomson CSF - DSD
Power Semiconductor Applications
50 Rue J P Timbaud
F-92403 Courbevoie
France

Mr. Cox
Eldurail B.V.
Postbus 132
NL-5530 AC Bladel
The Netherlands

Mr. D. Crees
Marconi Electronic Devices Ltd.
Carholme Road
UK-Lincoln LN1 1SG
England

Mr. P. de Bruyne
BBC Brown, Boveri & Co.
Dept. EKS
CH-5401 Baden
Switzerland

Dr. W.J. de Zeeuw
Techn. University Eindhoven
Department of Electrical Eng.
Den Dolech 2
NL-5600 MB Eindhoven
The Netherlands

Prof. Dr. Manfred Depenbrock
Ruhr-Universität Bochum
Universitätsstr. 150
D-4630 Bochum 1
West Germany

Mr. Ulrich Eckes
TCO-Traction CEM-Oerlikon
Avenue du Bel Air
F-69627 Villeurbanne
France

Mr. Peter Etter
BBC Brown, Boveri & Co.
Dept. IET
CH-5401 Baden
Switzerland

Prof. Dr. Willi Gerlach
Institut für Werkstoffe
der Elektrotechnik
Jebensstr. 1
D-1000 Berlin 12
West Germany

Dr. G. Gierse
Ruhr-Universität Bochum
Universitätsstr. 150
D-4630 Bochum 1
West Germany

Mr. H. Gilgen
BBC Brown, Boveri & Co.
Dept. KLM 13
CH-5401 Baden
Switzerland

Dipl.-Phys. Karl-Heinz Ginsbach
AEG-Telefunken
A51 (A513 L)
Postfach 2160
D-4788 Warstein 2
West Germany

Dr. Jens Gobrecht
BBC Brown, Boveri & Co.
Research Center
CH-5405 Baden
Switzerland

Dipl.-Ing. E. Göller
Institut für Elektrotechnik
an der Universität Stuttgart
D-7000 Stuttgart
West Germany

Dr. R.J. Grover
Mullard Hazel Grove
Bramhall Moor Lane
Hazel Grove
UK-Stockport, Cheshire SK7 5BJ
England

Mr. Claude Haglon
CEM-Division CEM-Systems-SCAM/FS
Department Redresseurs
de Puissance
40 Rue Jean Jaures
F-93176 Bagnolet Cedex
France

Mr. H. Hödle
BBC Brown, Boveri & Co.
Dept. I
CH-5401 Baden
Switzerland

Dr. Philip L. Hower
Unitrode Corporation
580 Pleasant Street
Watertown, MA. 02172
U.S.A.

Dr. Horst Irmler
Semikron GmbH
Postfach 820 215
D-8500 Nürnberg 82
West Germany

Dr. A. Jaecklin
BBC Brown, Boveri & Co.
Dept. EKS
CH-5401 Baden
Switzerland

Prof. Dr. Robert Jötten
Am Schlossgraben 1
D-6100 Darmstadt
West Germany

Dr. P. Kluge-Weiss
BBC Brown, Boveri & Co.
Research Center
CH-5405 Baden
Switzerland

Dipl.-Ing. Peter Knapp
BBC Brown, Boveri & Co.,
Dept. IES
CH-5401 Baden
Switzerland

Dr. Joachim Knobloch
Brown, Boveri & Cie. AG
Dept. HS/LE
Boveristrasse 1
D-6840 Lampertheim
West Germany

Dipl.-Ing. Karl Kommissari
Brown, Boveri & Cie. AG
Dept. VK
Postfach 3 51
D-6800 Mannheim
West Germany

Dr. M. Kurata
Toshiba R & D Center
Electron Devices Lab.
1, Komukai, Thoshiba-cho
Saiwai-ku
Kawasaki-City, Kanagawa 210
Japan

Ing. Lorenzo Lanzavecchia
T.I.B.B.
Piazzale Lodi 3
I-20135 Milano
Italy

Prof. Philippe Leturcq
Institut Nationale des
Sciences Appliquées
Avenue de Rangueil
F-31077 Toulouse Cedex
France

Ing. Karl-Heinz Lewalder
Kiepe Elektrik GmbH
Elektronik-Entwicklung
Thorner Strasse 1
D-4000 Düsseldorf 13
West Germany

Dr. A. Marek
BBC Brown, Boveri & Co.
Research Center
CH-5405 Baden,
Switzerland

Dipl.-Ing. Hans G. Matthes
AEG Elotherm GmbH
Hammesberger Str. 31
D-5630 Remscheid 1
West Germany

Dr. Rudolf Meier
BBC Brown, Boveri & Co.
Research Center
CH-5405 Baden
Switzerland

Prof. Dr. Hans Melchior
ETH Zürich
Inst. f. Angewandte Physik
& Abt. f. Industrielle Forschung
Häldeliweg 15
CH-8044 Zürich
Switzerland

Mr. Metzger
Heemaf B.V.
Postbus 4
NL-7550 GA Hengelo
The Netherlands

Prof. Dr. M. Michel
Techn. Universität Berlin
Institut f. Allgem. Elektrotechnik
Einsteinufer 19
D-1000 Berlin 10
West Germany

Prof. Dr. Gottfried Möltgen
Ebracher Weg 1
D-8520 Erlangen
West Germany

Dr. Arno Neidig
Brown, Boveri & Cie. AG
Dept. HS
Boveristr. 1
D-6840 Lampertheim
West Germany

Dipl.-Ing. Reinhold Neugebauer
Kiepe Elektrik GmbH
System-Engineering
Thorner Strasse 1
D-4000 Düsseldorf 13
West Germany

Prof. Dr. J. Nishizawa
Research Institut of
Electrical Communication
Tohoku University
Sendai
Japan

Dr. Hubert Patalong
Siemens AG, F4 H4T5
Kurt-Floencke-Str. 18
D-8000 München 60
West Germany

Mr. R. Pezzani
Thomson CSF - DSD
50 Rue J P Timbaud
F-92403 Courbevoie
France

Dr. Eberhard Pflüger
R. Bosch
Abteilung K9/EEF
Postfach 300240
D-7000 Stuttgart 30
West Germany

Dr. Karl Platzöder
Siemens AG
Frankfurter Ring 152
D-8000 München 46
West Germany

Prof. K. Ragaller
BBC Brown, Boveri & Co.
Research Center
CH-5405 Baden
Switzerland

Dr. P. Roggwiller
BBC Brown, Boveri & Co.
Research Center
CH-5405 Baden
Switzerland

Mr. W. Roos
BBC Brown, Boveri & Co.
Dept. CH-2
CH-5401 Baden
Switzerland

El.-Ing. Ch. Ruetsch
BBC Brown, Boveri & Co.
Dept. IEE-L
CH-5401 Baden
Switzerland

Dipl.-Ing. Heinrich Scheibengraf
BBC Brown, Boveri & Co.
Dept. IER-T
CH-5401 Baden
Switzerland

Prof. R. Schnörr
BBC Brown, Boveri & Co.
Dept. KL
CH-5405 Baden
Switzerland

Prof. J.A. Schot
Techn. University Eindhoven
Department of Electrical Eng.
Den Dolech 2
NL-5600 MB Eindhoven
The Netherlands

Prof. Dr. D. Schröder
Universität Kaiserslautern
Elektrotechnik
Pfaffenbergstr. 11
D-6750 Kaiserslautern
West Germany

Dr. C. Schüler
BBC Brown, Boveri & Co.
Research Center
CH-5405 Baden
Switzerland

Dr. R. Schüpbach
BBC Brown, Boveri & Co.
Dept. KLS-4
CH-5401 Baden
Switzerland

Prof. Dr. Francise Schwarz
Power Electronics Lab.
Dept. of Electrical Engineering
Delft University of Technology
NL-Delft
The Netherlands

Dr. Dieter Silber
AEG-Telefunken
Forschungsinstitut
Goldsteinstr. 235
D-6000 Frankfurt 71
West Germany

Dr. R. Sittig
BBC Brown, Boveri & Co.
Dept. EKS
CH-5401 Baden
Switzerland

Prof. Dr. H.-Ch. Skudelny
Inst. für Stromrichtertechnik
und Elektrische Antriebe
RWTH Aachen
Jägerstrasse 17-19
D-5100 Aachen
West Germany

Mr. John A. Slatter
Philips Research Laboratories
Redhill
UK-Surrey RHIA 5HA
England

Dr. Daniel T. Slattery
S.S.D. Ltd.
Eldon Way, Lineside
UK-Littlehampton Sussex
England

Dr. Karl Heinz Sommer
AEG-Telefunken
A51 (A513 L)
Postfach 2160
D-4788 Warstein 2
West Germany

Prof. A.P. Speiser
BBC Brown, Boveri & Co.
Research Center
CH-5405 Baden
Switzerland

Dr. H. Stemmler
BBC Brown, Boveri & Co.
Dept. IEE
CH-5401 Baden
Switzerland

Dr. Günter Syrbe
Brown, Boveri & Cie. AG
Dept. HS
Boveristrasse 1
D-6840 Lampertheim
West Germany

Mr. B.A. Tabak
Brown Boveri Nederland
Dept. VA
Postbus 301
NL-3000 AH Rotterdam
The Netherlands

Dr. Jeno Tihanyi
Siemens AG
Frankfurter Ring 152
D-8000 München 71
West Germany

Dr. Werner Tursky
Semikron GmbH
Postfach 820 215
D-8500 Nürnberg 82
West Germany

Mr. P. van Gelder
Smit Slikkerveer B.V.
Dept. Elektronika
Postbus 50
NL-2980 AB Ridderkerk
The Netherlands

Dr. Leonardo Vanotti
BBC Brown, Boveri & Co.
Dept. EK
CH-5401 Baden
Switzerland

Dr. Janis Vitins
BBC Brown, Boveri & Co.
Dept. EKS
CH-5401 Baden
Switzerland

Dipl.-Phys. B. Voss
Frauenhofer-Gesellschaft
Arbeitsgruppe für Solare Energiesysteme
Oltmannsstr. 22
D-7800 Freiburg
West Germany

Dr. Peter Voss
Siemens AG
Frankfurter Ring 152
D-8000 München 46
West Germany

Dr. J. Waldmeyer
BBC Brown, Boveri & Co.
Research Center
CH-5405 Baden
Switzerland

Mr. Hans H. Zander
Siemens AG
UB E GWE TS4
D-8520 Erlangen
West Germany

Dr. Rolf Zeyfang
AEG-Telefunken
Forschungsinstitut
Goldsteinstrasse 235
D-6000 Frankfurt 71
West Germany

Dr. Peter Zimmermann
Robert Bosch GmbH
GB Industrieausrüstung
Dept. EES 5
Postfach 12 49
D-6120 Erbach
West Germany

Dr. Wolfgang Zimmermann
BBC Brown, Boveri & Co.
Dept. EKS
CH-5401 Baden
Switzerland

Prof. Dr. Rudolf Zwicky
Inst. of Automatic Control and
Industrial Electronics
Swiss Federal Institute of Techn.
ETH-Zentrum
CH-8092 Zürich
Switzerland

HIGH-POWER INSTALLATIONS USING SEMICONDUCTOR DEVICES, INTERACTIONS BETWEEN SEMICONDUCTORS AND CONVERTERS

H. STEMMLER

BBC Brown, Boveri & Co. Ltd., Baden, Switzerland

SUMMARY

Power electronics has found a wide variety of applications in the fields of power generation, transmission and distribution, in industry and for traction drives. The
speed, direction, and extent of future progress in the development and application of
power electronics depend very strongly on the progress made in semiconductor technology, both regarding the signal and the power semiconductors. With new turn-off
thyristors, circuits with forced commutation, such as d.c. to a.c. voltage inverters,
and to d.c. choppers, may be drastically simplified. Circuits with line commutation,
belonging to the a.c. to d.c. voltage rectifier family, cannot be simplified in principle
but the characteristics can be improved. It may be assumed that the conventional nonturn-off thyristor has a long future and it is therefore worthwhile to invest considerable effort in its improvement.

1. INTRODUCTION

The task of power electronics is to provide means for the regulation of energy flow.
Accordingly a power electronic system always consists of a regulating and a power
electronic part linked together by the triggering electronics (Fig. 1). Either all or

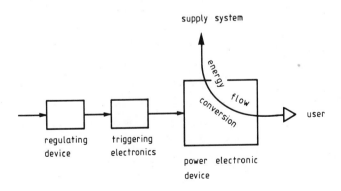

Fig. 1   Principle of a power-electronic system.

part of the energy to be influenced generally flows through the power electronics and thereby a conversion of the energy-carrying currents and voltages usually takes place into a form optimally suited to a particular user. Because of the variety of different applications, a corresponding variety of different regulating methods, triggering modes, and power-control devices with greatly differing characteristics has emerged covering power ranges ranging from watts to gigawatts.

Power electronic devices are used in many commercial areas, including energy generation, e.g. excitation systems and start-up frequency changers, energy transmission and distribution, e.g. HVDC, reactive power compensators, various industrial applications, and traction drives.

The significant advantages of power-electronic devices are that they are static and operate without wear, energy flows can be effected with practically no delay time, and continuing development will keep power electronics competitive pricewise in the future. The speed, direction, and extent of the progress in the development and application of power electronics depend very strongly on the progress made in semiconductor technology, both regarding the signal as well as the power semiconductors themselves. When considering the close connection between power semiconductors and power converters, consideration may be given to either the best use of components which are available today or beginning with the task to be solved and seeking an optimum circuit and power semiconductors with characteristics which are best suited to the task. In this paper the latter point of view is emphasized. Since the paper was presented as part of a scientific symposium, wishes are expressed which may not easily be fulfilled in practice.

In the following considerations high-power electronics (> 0.5 MW) are emphasized. A survey of important circuits and applications is presented in sections 2 and 3, and in section 4 some interactions between power semiconductors and power electronics are discussed.

2. THE MOST IMPORTANT CIRCUITS AND THEIR MAIN APPLICATIONS

In Figs. 2 and 3, the usual functions of present-day converters are listed, particular circuitry is named and typical applications given. In order to stay within the bounds of this paper, only an overview is presented without going into detail and deep explanations. Two of the most important basic circuits are discussed in somewhat more detail at a later stage in order to point out some areas of interaction between semiconductor elements and converter circuits.

When considering the function of converters it is necessary to distinguish between direct conversion, whereby one form of voltage or current is converted directly to another (section 2.1) and conversion in two steps with an intermediate d.c. link (section 2.2).

2.1 Circuits with Direct Conversion (Fig. 2)

It is necessary to distinguish between converters for which the impressed voltage on the input side is three-phase a.c. and those for which it is d.c.

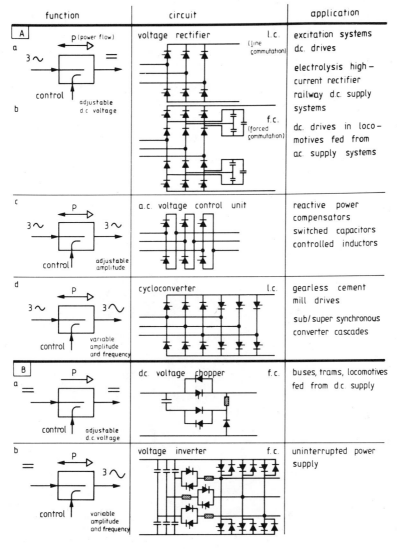

Fig. 2   Survey of the most important direct converters, their functions, circuits and main applications.

Circuits with an a.c. input include the following:

a) Voltage rectifiers with line commutation which produce an adjustable d.c. voltage. The most common circuit is the line-commutated controlled voltage rectifier bridge. Because of its simplicity this is one of the most widely used circuits, for example, to drive direct-current motors, as an electrolysis high-current rectifier, as a rectifier for railway supply systems.

b) Voltage rectifiers with additional forced commutation. Additional forced commutation gives the voltage rectifier the possibility of also operating on the a.c. side in both capacitive power quadrants. Such rectifiers are used in locomotives with direct-current traction motors fed from an a.c. supply system. In addition to their primary function as drive rectifiers they also fulfill the task of supporting the line by drawing capacitive current. The additional expenditure due to forced commutation is a barrier to their more widespread use at the present time.

c) a.c. voltage control unit which can either have the function of a circuit which usually switches on or off a three-phase a.c. voltage to a user, or can provide a.c. voltage of the same frequency with adjustable amplitude. The most important high-power applications are reactive power compensators in the form of switched capacitors and controlled inductances (Fig. 3).

d) Voltage direct frequency changers, (cycloconverters) which produce an a.c. voltage (usually three-phase) which is adjustable in amplitude as well as frequency and phase angle. Direct static-frequency changers (s.f.c.) consist of line-commu-

Fig. 3   a.c. voltage-control unit: one phase of a three-phase 450-MVA controlled reactor.

tated controlled voltage rectifier bridges. The output frequency must be limited to values below about 40 % of the frequency of the input voltage. An important application is the synchronous motor fed by a direct s.f.c. for gearless cement mill drives (Fig. 4). In addition, they are also used as sub-/supersynchronous converter cascades for the control of the transmission power of 50-16 $^2/_3$-Hz network-interconnecting converters for railways (Fig. 5).

Fig. 4   Gearless cement mill drive having 6 MW and 0-15 rpm.

Fig. 5   Rotating converters having 2 x 80 MVA for power transmission from 50-Hz supply system to 16 $^2/_3$-Hz single-phase traction supply system. Transmission power controlled by cycloconverters on the rotor side of the induction slipring motors.

Fig. 6   Streetcar with chopper-fed d.c. traction motors.

Direct converting devices with a constant d.c. voltage input are:

a) Choppers with forced commutation which produce a different adjustable d.c. voltage. These are used in particular in overhead contact d.c. voltage vehicles for driving direct-current traction motors (Fig. 6).

b) Voltage inverters which are also force-commutated and produce an a.c. voltage from constant d.c. voltage by means of pulse-duration modulation. The voltage is in general variable in amplitude, frequency, and phase angle. These are mainly used for the continous supply of a.c. voltge from batteries.

With the exception of the chopper, all the circuits mentioned have the characteristic that energy can be transported in the direction from the side of the impressed voltage to the side of the voltage produced, and vice versa. They can be used to build up circuits with an intermediate d.c. link, and so have functions which are not possible with direct conversion and thereby leading to further applications.

2.2 Circuits with Converters with Intermediate d.c. Link (Fig. 7)

Here again we have to distinguish between converters with a.c. and d.c. input.

Circuits using an intermediate d.c. link with an a.c. supply voltage are:

a) Frequency changers with a d.c. current intermediate link consisting of two line-commutated voltage rectifiers the d.c. sides of which are connected back-to-back by a reactor. Both external connections must be to a.c. voltage sources since

function	circuit	application
A a 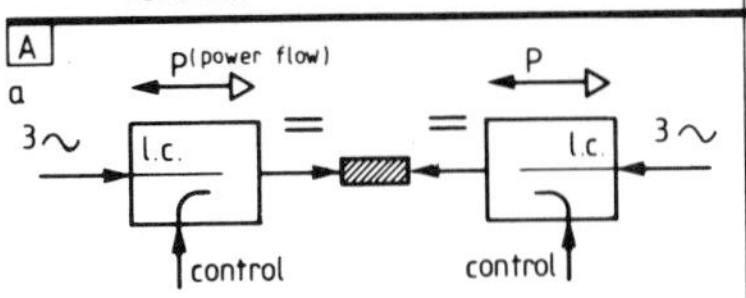 p(power flow) → P → 3~ [l.c.] = = [l.c.] 3~ control control	voltage rectifier l.c. (line commutation) d.c. current link voltage rectifier l.c.	HVDC transmission large drives with synchronous machines start-up equipment inductive heating and melting
b 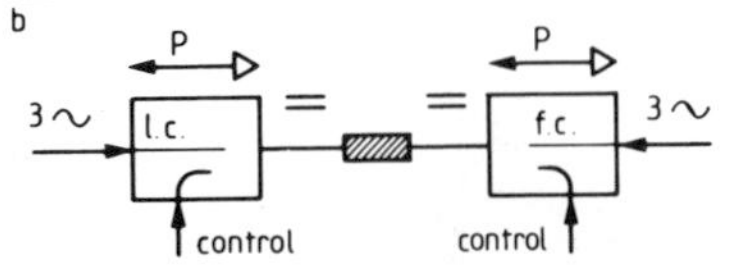 P → P → 3~ [l.c.] = = [f.c.] 3~ control control	voltage rectifier l.c. d.c. current link voltage rectifier f.c. (forced commutation)	a.c. drive with induction motor
c 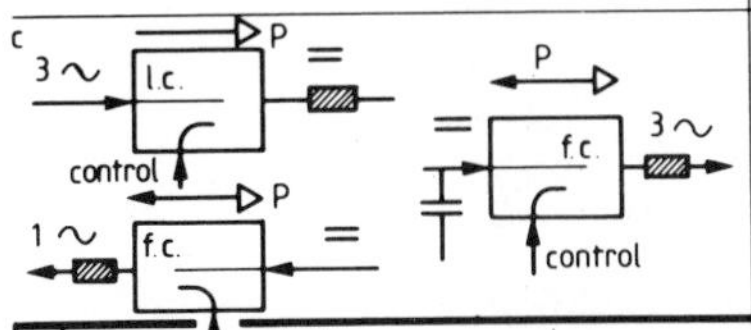 3~ [l.c.] → P control = = [f.c.] 3~ 1~ [f.c.] = control	voltage rectifier l.c. voltage inverter f.c. d.c. current / voltage link d.c. voltage link voltage inverter f.c.	diesel-electric / overhead-contact wire locomotives with induction motor drive
B a 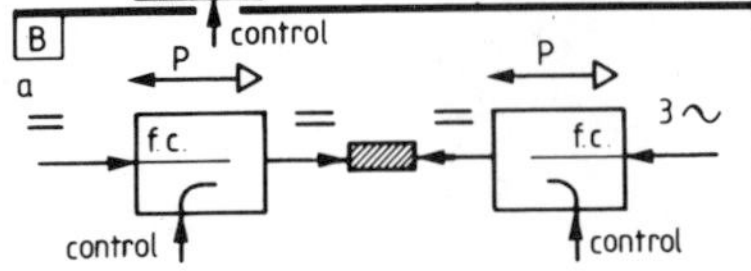 control P → P → = [f.c.] = = [f.c.] 3~ control control	d.c. voltage chopper f.c. d.c. current link voltage rectifier f.c.	buses / trams fed from d.c. supply voltage

Fig. 7   Survey of the most important converters with an intermediate d.c. link, their functions, circuits and main applications.

Fig. 8a   The power electronic towers 500 MW, 250 kV of a high-voltage d.c. transmission.

these devices are not able to generate a.c. voltage themselves. The most important
high-power application of this principle is high-voltage d.c. transmission (HVDC)
for the coupling of three-phase current networks over long distances (Fig. 8).
This principle is also useful for three-phase drive systems if the motor is a synchronous motor. Such drives have been built for power up to 30 MW (Fig. 9) and

Fig. 8b   Thyristor module of a HVDC-converter.

Fig. 9   30 MW-rectifier for a large synchronous machine drive.

Fig. 10   Normalized start-up devices for gas turbine generator

have found application particularly, but not exclusively, as start-up devices for gas-turbine generators (Fig. 10). As a variant of this system there is another important application: inductive heating or melting with medium frequencies of some 100 or 1000 Hz. In this case capacitors are added to the charge at the user side in order to allow commutation without additional commutation circuits.

b) Frequency changers with a d.c. current intermediate link consisting of two voltage rectifiers the d.c. sides of which are connected back-to-back by way of a reactor. The principle is closely related to the one mentioned previously. Now, however, the supply side is line-commutated and the machine side is forced commutated. The machine side with forced commutation is necessary if a squirrel-cage motor is to be used. For line-commutated converters, squirrel-cage motors are not suitable voltage sources because they cannot absorb capacitive current (that is, they cannot produce inductive current).

c) Frequency changers with d.c. current voltage or a pure d.c. voltage intermediate link with a force-commutated voltage inverter on the motor side for the production of a voltage variable in amplitude and frequency with which the motor is loaded. This principle is usually used for squirrel-cage motors.

On the line side there are two variants. In the case of the current voltage intermediate link, a line-commutated voltage rectifier is used. For a pure d.c. voltage link, a voltage inverter with forced commutation controlled by pulse-duration modulation is also used on the supply side. Thus almost ideal conditions result for the supplying network which supplies pure active current with a very good sinusoidal waveform.

The two principles mentioned have found application principally in the area of traction: the former for diesel-electric three-phase locomotives (Fig. 11) the latter for 16 $^2/_3$-Hz overhead-contact wire locomotives with three-phase drive.

Fig. 11   Diesel-electric locomotive with inverter-fed squirrel-cage motors.

Converters with intermediate d.c. link with direct current supply voltage are:
a) Converters with d.c. current intermediate link with a chopper on the supply side and voltage rectifier on the output side which operate with forced commutation in order to drive a squirrel-cage motor. This principle is gradually being introduced at the present time particularly for short-distance traction vehicles, such as overhead-contact-wire vehicles which are fed from a d.c. voltage supply.

It was the purpose of the foregoing to give a rough overview of the spectrum of functions, circuits and important applications. However, a systematic discussion of the wide variety of circuits just mentioned and the demands they place on power semiconductors would go far beyond the scope of the present paper. Nevertheless, it can be shown that this wide variety of circuits can in principle be derived from two basic types of circuit which can occur in different variations and combinations:

　　- the (generally) line-commutated a.c. to d.c. voltage rectifier
and
　　- the d.c. to a.c. voltage inverter with forced commutation.

These circuits will now be discussed in further detail from the following points of view:
- significant characteristics (section 3)
- the simplifications and improvements to be expected from new semiconductors with quantitatively and qualitatively new characteristics (section 4)

3. MAIN CHARACTERISTICS OF THE TWO MOST IMPORTANT CIRCUITS: THE D.C. TO A.C. VOLTAGE INVERTER AND THE A.C. TO D.C. VOLTAGE RECTIFIER

3.1 The d.c. to a.c. Voltage Inverter with Forced Commutation (Fig. 12)

This circuit has a d.c. voltage impressed on the input side from which it produces an a.c. voltage on the output by means of control. Functionally the circuit principle is very simple consisting (per phase) of a change-over switch which can switch its output either to the positive or negative pole of the d.c. voltage supply. The method of pulse-width modulation permits the adjustment of the amplitude with respect to the frequency of the generated a.c. voltage. To put this simple principle into practice with today's power semiconductors is no longer simple however. In addition to the two main thyristors and associated antiparallel diodes, a commutating device is needed consisting of commutating capacitors, reactors and two additional commutating thyristors. It may be seen that the voltage inverter switches the voltage on the a.c. voltage side and the current on the d.c. voltage side.

Two limiting conditions result, which must be met with this type of circuit. First, it is only possible to switch currents if they are conducted to or from a constant-level voltage source. This implies that the impressed d.c. voltage must be supported by capacitors. Such a voltage inverter could never be connected to a constant-level stable current source. Secondly, the output of the inverter can never be connected to a constant-level external voltage source, because voltage switching would otherwise be impossible.

In principle the moment of triggering of a voltage inverter can be chosen without limitation. Switching can therefore occur at any current polarity to any voltage polarity on the a.c. voltage side, and this results in the fact that the device operates on the a.c. voltage side freely in all four power quadrants.

In the case of faulty commutation the impressed d.c. supply can be short circuited via the voltage inverter. In order to avoid destruction of the thyristors, the inverter needs a very fast and efficient protection device, consisting of one or more large short-circuit thyristors, which takes up the discharge current from the backup capacitors.

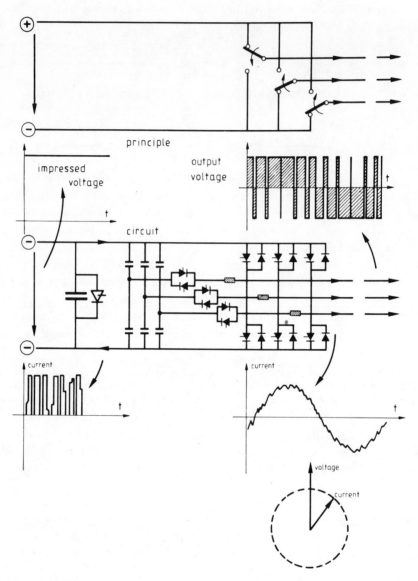

Fig. 12   Principle and circuit of d.c. to a.c. voltage inverter with forced commutation.

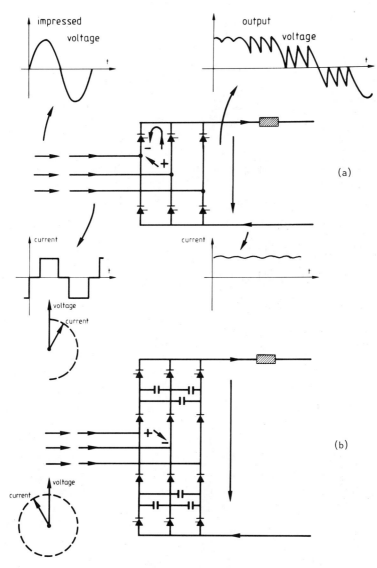

Fig. 13 a.c. to d.c. voltage rectifier with line commutation (a) and with forced commutation (b).

3.2 The a.c. to d.c. Voltage Rectifier with Line or Forced Commutation (Fig. 13)

This circuit has an a.c. voltage impressed on the input side from which it produces a d.c. voltage on the output side by means of control. The circuit can be built up with only two thyristors per phase. The generated d.c. voltage can be adjusted between a positive and a negative maximum value by adjusting the triggering times of the thyristors with respect to the a.c. voltage supply. The direct current, on the other hand, can only flow in one direction, which is given by the direction of the thyristors. Despite the given current direction, reversing the energy flow is possible by reversing the generated d.c. voltage.

The practical realization of this circuit with today's power semiconductors is already so simple that no further fundamental circuit simplifications can be expected as a result of future power semiconductors with qualitatively new features. Nevertheless, for many applications this circuit will benefit from quantitative improvements in the thyristors. It may be seen that the a.c. to d.c. voltage rectifier switches the voltage on the d.c. output side and the current on the a.c. input side. Two limiting conditions therefore result which must be met with this circuit. First, the a.c. supply voltage must be as stable as possible in order to make current switching possible. Secondly, the output of the voltage rectifier may on no account be connected to a constant-level d.c. voltage source because then voltage switching would be impossible. At the output there is therefore usually some form of inductance, even if it is only the leakage inductance of a d.c. motor.

The moment of triggering which starts the current commutation from one thyristor to the next can only be chosen within certain limits, corresponding to the two half periods in which the interlinked a.c. supply voltage has the polarity to open the newly fired thyristor and close the previously conducting thyristor. A consequence is that the voltage rectifier can operate on the a.c. voltage side in only two of the four power quadrants, i.e. those in which it has an inductive component as a load, this being sufficient to reverse the direction of the energy flow but not to draw capacitive current. This limitation is something of a disadavantage. However, a variant of the voltage rectifier can be used to make the capacitive regions of the four power quadrants also accessible (Fig. 13b). For this purpose the voltage rectifier is provided with additional forced commutation whose commutating capacitor provides the voltage polarity necessary for commutation, even in those periods when the a.c. supply line voltage has the "incorrect" sign.

The advantages of this circuit are obvious. First, in cases where the voltage rectifier with forced commutation is located on the line side, it can also be operated in a capacitive, line-supportive manner. Secondly, with drives where it is located on the machine side, a squirrel-cage motor can also be driven which, as a load, can only operate in the inductive power quadrants, i.e. delivering capacitive current to the voltage rectifier. The obvious disadvantage of this circuit is its higher complexity.

4. THE CONTRIBUTION OF SEMICONDUCTOR DEVELOPMENT TO THE
SIMPLIFICATION AND IMPROVEMENT OF INVERTERS AND RECTIFIERS

In this section we discuss how semiconductor development can contribute to the simplification of converters and/or the improvement of their characteristics.

4.1. d.c. to a.c. Voltage Inverters (Fig. 14)

One of the negative characteristics of present-day thyristors, which is distinctly disadvantageous here, is that it can only be switched on but not switched off. The

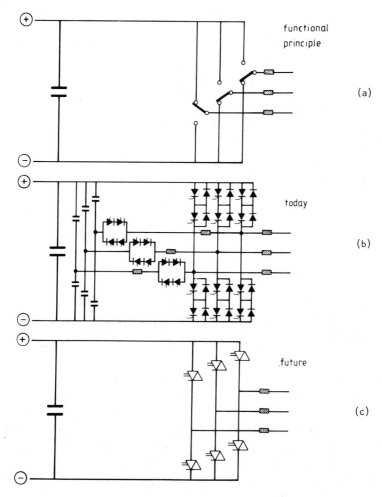

Fig. 14   d.c. to a.c. voltage inverter: (a) functional principle, (b), present circuit,
            (c) possible circuit of the future.

capability for turn-off, which is required by the circuit but lacking in the thyristor,
can only be realized today with considerable additional equipment outlay (Fig. 14b).
This additional equipment outlay arises in part from the fact that it is necessary to
provide commutation circuits with thyristors, capacitors, and chokes. Also, frequency
thyristors are involved, and these are more expensive than line thyristors (Fig. 15).
Finally, these thyristors are available for voltages of only about half those for line
thyristors. Thus already in the case of average power below 1 MW a series connection
of thyristors becomes necessary thereby further adding to the expense.

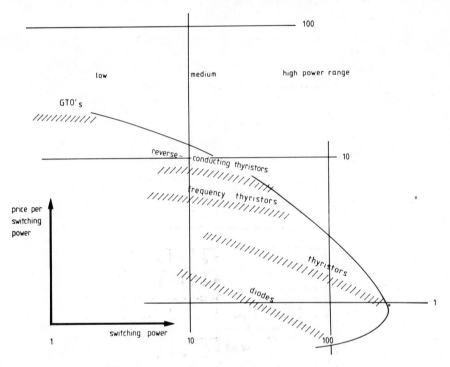

Fig. 15   Price per switching power as a function of switching power (repetitive
            blocking voltage x surge current) for different power semiconductors:
            diodes, normal thyristors, frequency thyristors, reverse-conducting
            thyristors and gate turn-off thyristors.

This situation can be improved by the use of frequency thyristors with a higher
blocking voltage. Since the main thyristors in most voltage-inverter circuits are only
stressed voltage-wise in the blocking direction, a "one-sided" increase in the blocking
voltage would be useful. It is also possible to use frequency thyristors with integrated antiparallel diodes (RCTs) whereby space for installations in the converter
can be saved. The use of turn-off thyristors allows elimination of the entire commutation circuit. A further step in reducing the outlay would be turn-off thyristors with
integrated antiparallel diodes.

Fig. 14 shows the comparison of d.c. to a.c. voltage inverter both with today's power elements (Fig. 14b) and how it might look in future (with reverse-conducting turn-off thyristors) (Fig. 14c). Even from this very qualitative viewpoint, the simplification resulting from only one component per function is evident.

Fig. 14 shows the possibilities of circuit simplification for inverters. It remains to be considered which characteristics can be achieved using such inverter circuits. For purposes of illustration we take the example of an inverter locomotive with overhead-contact wire fed by the 16 $^2/_3$-Hz supply system (Fig. 16).

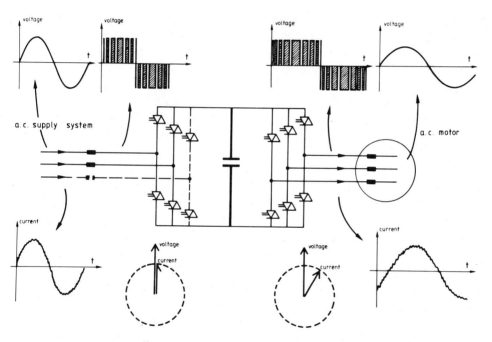

Fig. 16   a.c. to a.c. converter with intermediate d.c. voltage link based on future circuit of Fig. 14(c).

Both on the user side and the line side, the circuit consists of voltage inverters with forced commutation, the d.c. voltage sides of which are capacitor-supported and connected back-to-back. The a.c. voltage sides are connected on one side to the induction motors and on the other by way of a reactor or transformer leakage inductance to the a.c. supply system. Voltage with the necessary frequency and amplitude variation is impressed by the motor-voltage inverter. By using appropriate pulse-duration modulation, an almost sinusoidal motor current can be achieved. In the same way, the voltage inverter itself produces an a.c. voltage on the supply side which at the same frequency can, for example, be adjusted as desired in amplitude and phase angle relative to the vector of the 16 $^2/_3$-Hz supply voltage. As a result, pure active current can be drawn from the railway power supply with a waveform which here again can be controlled with correspondingly fine pulse-duration modulation to achieve an almost pure sinusoidal form.

In this way an approach is made towards the ideal inverter which is harmonic-free, both on the user side and the supply side, which can draw from the a.c. supply system either pure active power or for compensation purposes, capacitive or inductive reactive power, and which can deliver to the user side any desired active or reactive power. These advantages cannot be achieved by any other circuit today.

The application of this circuit is, of course, in principle not limited to locomotives. With voltage inverters, and the related choppers, as well as with combinations of such basic circuits, all necessary functions can be achieved: a.c. to d.c., a.c. to a.c., d.c. to d.c. and d.c. to a.c. The question may therefore be posed as to whether voltage inverters and related circuits will, to a large extent, replace other circuits of the voltage rectifier family. A related question is whether, as has been suggested, turn-off semiconductors will completely replace nonturn-off thyristors. In attempting to answer these questions, two points shall be considered. First whether voltage-inverter circuits have limiting disadvantages as well as distinct advantages. Secondly, whether voltage-rectifier circuits could also benefit from turn-off semiconductors.

So far as the first question is concerned, even after exhausting all possible simplifications through new types of semiconductors, some limiting characteristics will probably still remain. Just as today, the cost of frequency thyristors is higher than that of line thyristors, realistically it must be assumed that the cost of more sophisticated semiconductors will be higher in future (Fig. 15). The necessary stable, constant-level, and therefore capacitor-supported, voltage on the d.c. side requires fast and efficient protection to prevent high short-circuit currents from flowing through the inverter. A strong reduction in the harmonics content is gained by an increasing switching frequency (e.g. 10 or 100 switches per half cycle of the generated voltage). Compared to converters, which switch only once per half cycle, the high frequencies will probably always involve direct or indirect additional expenditure.

4.2 a.c. to d.c. Voltage Rectifier (Fig. 17)

The second question posed above was whether circuits which belong to the voltage-rectifier family might also benefit from semiconductors with qualitatively new characteristics. Voltage rectifiers in contrast to voltage inverters are already so simple in their circuitry with present-day thyristors, that in principle they cannot be further simplified even with new types of semiconductors (Fig. 17a). Nevertheless, they do have characteristics which could still use improvement. The current drawn from the line on the a.c. voltage side has a rectangular waveform and therefore contains harmonic components. In addition, the current becomes increasingly more inductive with decreasing degrees of modulation.

It may be recalled that the a.c. to d.c. voltage rectifier switches the current on the a.c. side. It is then clear that turn-off semiconductors cannot be used quite so

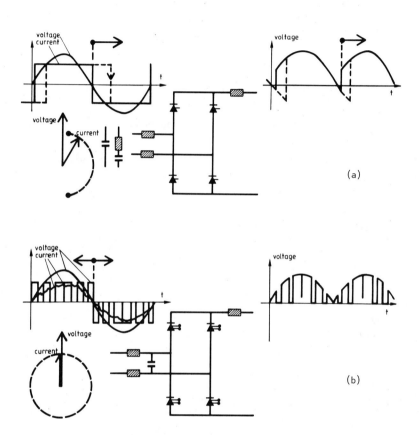

Fig. 17   a.c. to d.c. rectifier: (a) present-day circuit and function, (b) possible circuit and function of the future.

simply (Fig. 17b): the internal inductance, which is almost always present even for stable, constant-level supply systems, would resist the switching off of the current and react with overvoltages which would damage the semiconductors. In an analogous way to voltage inverters, a solution is capacitor support of the line a.c. supply voltage and chopping the current into pulses with sinusoidal pulse-duration modulation. The higher the chopper frequency, the less expenditure is necessary for support capacitors to prevent current-switching overvoltages. Fig. 17b shows clearly how the characteristics of the circuit can be improved in this way. The current drawn from the line becomes more and more sinusoidal with increasing chopper frequency, and the current can be regulated so that cos ϕ = 1, or it can have an inductive or capacitive component.

It is still difficult to predict whether these improvements in circuit characteristics will be more economical by using turn-off elements or normal nonturn-off thyristors plus harmonic filters and compensating capacitors, as is the case today. In the long term it will probably be cheaper to replace as many copper windings, iron cores, and aluminum capacitor wrappings as possible with semiconductors. In any case it must be assumed that the conventional present-day nonturn-off thyristor has a long future and it is still well worthwhile investing considerable effort in its improvement.

4.3 Improvement of the Conventional Nonturn-Off Thyristor

4.3.1 Simplification of the Trigger Circuitry

The triggering of thyristors today still involves relatively high outlay.

a)   Converters in the lower and medium power ranges (Fig. 18) require trigger-pulse amplifiers together with their supply, isolating trigger-pulse transformers and careful planning of the grounding, zero-voltage, and shielding requirements. It

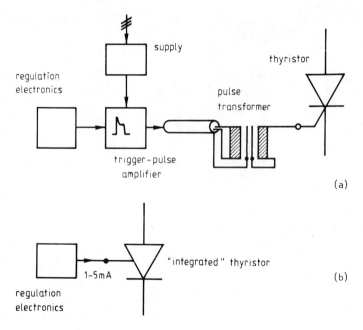

Fig. 18   Principle of trigger circuit for low- and medium-power converters: (a) actual circuit, (b) possible circuit of the future.

may be that these requirements could be replaced by an "integrated" thyristor triggered by a few milliamps, which incorporates a potential separation, e.g. using light, and which draws the energy for the gate impulse from the anode-cathode voltage. Such a solution might be possible using a combination of V-MOS and thyristor technology.

b) In the area of high-power electronics, thyristor triggering is also still relatively complex and expensive (Fig. 19a). In the present-day technique, the potential barrier between the control devices and the thyristor is overcome by light guides and each thyristor requires closely associated electronics which receives and converts the light signals, checks the status of the thyristor (V, dV/dt, BOD-firing), takes energy out of the RC network, fires the thyristor, and feeds back the status via light guides.

Here again significant simplifications are possible by triggering the thyristor or an associated auxiliary trigger thyristor directly with light. Although such light-activated thyristors are available, they need much more energy than can be provided by today's LEDs (30 to 300 µs and 5 mW instead of 30 µW). Improvements could be made on two fronts; first, using more powerful LEDs with better light coupling in the light guides, and secondly with thyristors that are more light (but not interference) sensitive.

4.3.2 Thyristor with High Power

In general, converter circuits can be realized most economically if it is possible to use only one simple element per function. In high-voltage converters a series connection of very many thyristors per bridge arm will still be necessary. However, thyristors with even higher blocking voltages than the ca. 4.5-kV elements of today would further reduce the cost of converters because thyristors with higher power have a lower cost per power rating, and (Fig. 15) converter design and triggering become less expensive if the number of built-in components is decreased. Thus elements with a higher blocking voltage are clearly desirable.

High-voltage applications are already realized today practically without parallel switching of thyristors, so that higher voltage is clearly preferable to higher current. However, for protection purposes, it is desirable to have the highest possible short-circuit current, permitted once or a few times per half cycle.

4.3.3 Reduction of Power Losses

Increasing emphasis is now being given to power-loss evaluation. The costs lie in the region of several 1000 Sfr. per kW power loss and, for extra-high-power applications,

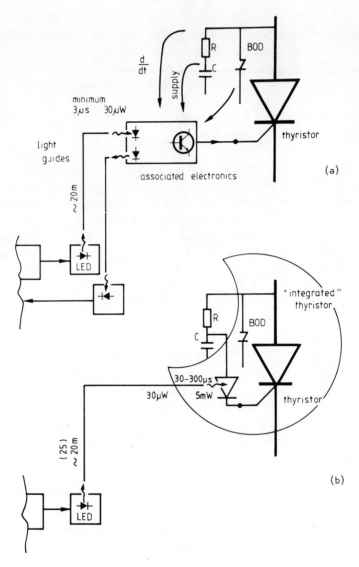

Fig. 19   Principle of trigger circuit for high voltage power converters. Actual circuit (a). Possible circuit in future (b).

reach an order of magnitude which is already significant in comparison to converter costs. Decreasing thyristor losses therefore has a high priority. At the same time attention must be paid to the fact that the total losses of a converter consist roughly of about 50 % thyristor losses, given by the forward-voltage drop, and 50 % RC-network losses, which depend on reverse-recovery charge. Since forward-voltage drop and reverse-recovery charge are in a reverse interdependence, a simple reduction of the forward-voltage drop would only shift the thyristor losses to the RC network.

DISCUSSION
(Chairman: Dr. W. Gerlach, TU, Berlin)

F. Schwarz (Delft University of Technology, Delft)

As I understand it from the presentation the mechanics of turning off thyristors would be greatly simplified by the use of GTO's. Are there any data available that show how a reduction in the losses of a switch might be achieved? I refer to the losses that are caused by the storage of energy in the inductive elements that are in series with the switch that breaks the circuit, and also the power needed in order to generate the pulses that interrupt the current.

H. Stemmler (BBC Brown, Boveri & Co. Ltd., Baden)

We have as yet no experience with practical applications of the gate turn-off thyristors because they have such low power that they cannot be used in high-power electronics. We are now beginning experiments with low-power electronic devices which have to switch with high frequency. We will then evaluate as to whether we should use bipolar transistors or power MOS or gate turn-off thyristors.

H. Irmler (Semikron GmbH, Nürnberg)

Mr. Stemmler told us that it would be very interesting to have big power thyristors with voltages above 5000 V. What are your thoughts regarding even higher voltages, for instance for HVDC applications and so on where there are many problems to be overcome?

H. Stemmler

Our aim is for high-voltage thyristors because with such thyristors we could reduce costs in two ways. One is that with increasing switching power the price drops. The second point is that if you have less elements to install in the converter you have lower installation costs and you need less firing and triggering elements.

R. Jötten (TU, Darmstadt)

You pointed out quite clearly the relation between switching losses and expenditure in connection with pulsewidth modulation. However there is also a very important relation with switching frequency. You said this should be as high as possible, but losses also increase with the switching frequency. Have you investigated this problem systematically?

H. Stemmler

Since we do not yet have high-power, high-frequency thyristors we cannot start such investigations. At the same time, inductive heating applications with powers of about 2 MW are realized with frequencies of 1 kHz or more and only the future will show if it is less expensive to use higher frequency thyristors or to use normal thyristors with additional harmonic filters and capacitive compensators.

R. Jötten

Have you considered the losses in the auxiliary circuitry as well as in the main circuit?

H. Stemmler

Losses are included in the price because they are evaluated by the client, and as I said this morning, especially in very high power installations the cost of power losses reaches the same magnitude as the hardware of a compensator, for example.

HIGH-POWER THYRISTORS

A.A. JAECKLIN
BBC Brown, Boveri & Co. Ltd., Baden, Switzerland

SUMMARY

The field of high-power thyristors is reviewed from a historical point of view. In terms of switching power, it is found that there has been substantial growth over the years. The various properties of high-power thyristors are treated with respect to their influence on the electrical parameters. Most important for the static behavior are neutron-transmutation doping of the starting material and the beveling geometry of the semiconductor edge. In contrast, the dynamic behavior is dominated by the turn-on properties leading to a distributed gate structure and the turn-off behavior which requires careful control of the recovered charge, Q_{rr}. Thus the tradeoff is very similar to that for a fast-switching thyristor. As far as future developments are concerned, it is expected that most emphasis will be placed on a further increase of blocking voltage and that switching power will grow at a slower pace than hitherto.

1. INTRODUCTION

It is the purpose of this paper to review the scope and state of the art of high-power thyristors in a rather general way and to cite specific results only where data are needed for purposes of illustration. The classical field of power semiconductors is considered here, operating with standard line frequency (i.e. 50 or 60 Hz). Emphasis will be placed upon the high-power end of the spectrum. Obviously, this segment of applications involves self-commutating circuits and phase-controlled sinewave currents. After brief consideration of the evolution of power thyristors, the influence of the most important static and dynamic parameters will be considered. On this basis the present state of the art will be discussed as it has evolved in close cooperation with specific important applications, such as high-voltage d.c. transmission (HVDCT).

2. EVOLUTION

Historically, high-power d.c. motors were the first major application of power thyristors. In the mid sixties, the first high-voltage valves were built on an experimental basis, competing at that time with mercury vapor thyratrons.[1] Subsequently work on

27

practical HVDCT was started, leading to working systems in the early seventies.[2] An additional important application was added in 1975 with the static compensation of reactive power.

The evolution of high-power semiconductor elements has always been limited by fundamental problems originating from materials quality and from the process technology. Considering the fact that a given silicon wafer yields only one device, not even a single defect is permitted. This concept of zero defects may be illustrated by comparing the wafer size for early power devices of about 3 cm^2 (20-mm dia.) with present-day wafers of approximately 75 cm^2 (100-mm dia.). In order to maintain a constant yield, the defect density has evidently been reduced by a factor of 25. Additional improvements are required as is illustrated in Fig. 1a where a 100-mm dia. wafer exhibits significant crystal damage after a process sequence optimized for wafers of half that diameter. This damage can be avoided with correctly adapted processes as in Fig. 1b.

a   b

Fig. 1   Process-induced crystal damage after Sirtl etch of 100-mm dia. silicon wafers:
a)   process sequence optimized for 50.8-mm dia.
b)   processes adapted to 100-mm dia.

Handling power is identical to keeping thermal effects under control. In contrast to other approaches,[3] the maximum repetitive load current I_{TM} with a 180° conduction angle at a fixed case temperature (80°C) is chosen in combination with the rated repetitive blocking voltage V_{DRM} as shown in Fig. 2. The switching power P_s is thus defined by the instantaneous peak value:

$$P_s = V_{DRM} \cdot I_{TM} \tag{1}$$

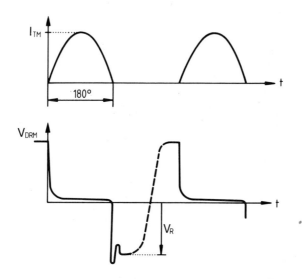

Fig. 2   The maximum repetitive current I_{TM} at a given case temperature (T_C = 80°C) and the repetitive voltage V_{DRM} used to define the switching power $P_s = V_{DRM} \cdot I_{TM}$.

An estimate for P_s can be extracted from most data sheets and Fig. 3 illustrates the approximate development of power thyristors over the years. An uncertainty stems from the time of introduction for a new component which lies somewhere between feasibility and large-scale application. It is interesting to note that except for a certain time lag, the various kinks in the development of V_{DRM} can be traced to specific advances in the state of the art, as for example:

1960	influence of surface beveling[4]
1964	first quantitative model of bevel angle[5]
1974	neutron-transmutation doping of silicon[6]
1974/75	novel beveling geometry; quantitative influence of surface charges[7],[8]

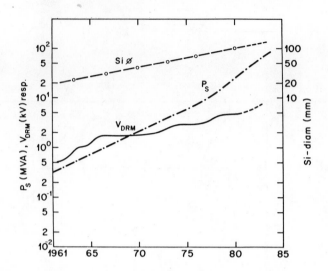

Fig. 3   Evolution of maximum applied voltage (V_{DRM}) wafer size (Si ϕ) and switching power (P_s) for high-power thyristors. The time required for doubling
P_s has decreased from three years to less than two.

The introduction of successively increasing wafer sizes of suitable quality (i.e. float
zone Si) is indicated by the circles in Fig. 3. As shown by the dashed line, the
growth of wafer size can be fairly well approximated by an exponential. The current-
carrying capability increases almost proportionately to the area of wafer and has been
omitted from Fig. 3 in the interest of clarity.

The resulting switching power P_s has shown a steady increase, on average doubling in
size every two and a half years. It should be noted that this growth compares very
reasonably to the increase of complexity on an integrated circuit (e.g. the storage
capacity on a memory chip[9]). High-power thyristors which represent the state of the
art are currently produced both in Europe[3,10] and in Japan.[11,12]. The more recent
devices have current-carrying capabilities high enough (I_{TAVM} about 2000 A) that
parallel-connected devices are no longer needed for the highest power that is current-

ly handled by a single system. This indicates that the useful limits of switching power may be approached in the not too distant future.

3. STATIC BEHAVIOR

The high-power thyristor will be discussed pragmatically by reviewing its physical properties from the point of view of its influence on electrical performance. For high-power devices, the static behavior is fundamental since the static losses are of principal importance to the efficiency of the device. It is of importance that the guidelines for device design are worked out carefully from the point of view of the main application in order to optimize the system as a whole.

3.1 Silicon Material

The highest voltages ever attempted in semiconductor technology applied in most cases to high-power thyristors. With increasing breakdown voltage V_B, the doping N_D of the starting material decreases almost in inverse proportion:[13]

$$V_B \propto \frac{1}{N_D^{3/4}} \qquad (2)$$

At high blocking voltages, the required doping becomes very small (e.g. 2×10^{13} cm^{-3}). Hence local doping variations of conventionally doped silicon as they are induced by the crystallization process[14] become unacceptable. A real breakthrough has been achieved by the use of neutron-transmutation doping (NTD).[6] As shown in Fig. 4, there are three isotopes of silicon, the natural distribution of which is given in the last column. Upon irradiation with thermal neutrons, various reactions occur of which only the one involving ^{30}Si is of importance because it leads to a phosphorus doping. It may be noted that ^{31}Si has a conveniently short half life of 2.6 hours. The secondary reaction involving radioactive ^{32}P shown at the bottom of Fig. 4, however, may lead to an increase in storage time until the radiation associated with it has decayed to a safe level.

Because the neutron penetration depth is larger than presently used crystal diameters, a very homogenous distribution of the doping results, being limited primarily by the lateral profile of the neutron flux itself (e.g. $\Delta\rho/\rho = 2$ % for $\rho = 200$ $\Omega.cm$). The improvement in homogeneity over conventionally doped high-resistivity material is more than a factor of ten.[15]

distribution

$$^{28}Si\ (n,\gamma) \longrightarrow {}^{29}Si \qquad\qquad 92,3\%$$

$$^{29}Si\ (n,\gamma) \longrightarrow {}^{30}Si \qquad\qquad 4,7\%$$

$$^{30}Si\ (n,\gamma) \longrightarrow {}^{31}Si \xrightarrow{2,6h} {}^{31}P + \beta \qquad\qquad 3,1\%$$

$$^{31}P\ (n,\gamma) \longrightarrow {}^{32}P \xrightarrow{14,3d} {}^{32}S + \beta$$

Fig. 4   Nuclear reactions triggered by irradiation of silicon with thermal neutrons. Only the reaction involving ^{30}Si leads to the desired doping. The figures on the right-hand side give the natural distribution of the silicon isotopes.

Since fast neutrons produce not only additional unwanted activity but also create additional crystal damage, irradiation by slow thermal neutrons is preferred as it can be reasonably well approximated by a swimming pool reactor. Comparing the energy released by the individual processes to the energy required to displace a single silicon atom from its site, an estimate of the crystal damage can be obtained.[16] These processes, as listed in Fig. 5, are dominated by "fast neutrons" involving all the damage caused by neutrons until they are slowed down such that they can be captured by a silicon atom.

pool reactor

fast neutron	$1,4 \times 10^4$ displaced Si
fission gamma	$3,6 \times 10^1$
gamma recoil	$1,3 \times 10^3$
beta recoil	$2,8 \times 10^0$
total	$1,5 \times 10^4$

Fig. 5   Crystal damage caused while creating a single phosphorus atom by neutron-transmutation doping (calculated values).

Comparatively less damage results from fission and recoil from gamma and beta particles. It must be regarded as a gift of nature that the subsequent high-temperature annealing succeeds in healing the enormous amount of crystal defects so well that apparent effects in device behavior are no longer noticeable.

3.2 High-Voltage Passivation

The magnitude of the relative dielectric strength of silicon ($\varepsilon = 12$) may lead to excessively high electric fields immediately outside a reverse-biased p-n junction in the area where the SCL reaches the surface. It is well known that planar junctions give satisfactory blocking capability only up to several 100 V while higher voltages need some kind of surface contouring in order to reduce the local electric field in the adjacent material.

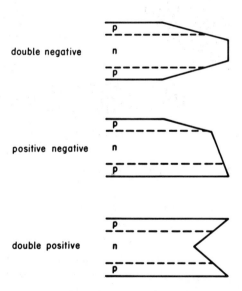

Fig. 6   The most common surface contours for high-power thyristors.

The most important contours used for high-power thyristors are shown in Fig. 6. Although the double-negative bevel has the advantages of easy machining and of symmetry, a field maximum occurs a short distance below the surface[17] which limits the voltage to a value below that of the one-dimensional structure. With the positive-negative bevel which is most commonly used for alloyed devices, the blocking capability of the lower p-n junction can be fully utilized; unfortunately there are only few applications where this is a real benefit.

Optimum design is possible with the double-positive bevel where full advantage of the potential blocking voltage can be taken.[7] This in turn means lower forward losses for a given off-state behavior. The disadvantage is more difficult machining of the contour as well as higher surface fields in most cases. A practical example of such a contour is shown in Fig. 7. This contour must of course be protected by some inert material that helps to avoid the accumulation of charges on the surface (typically silicon rubber).

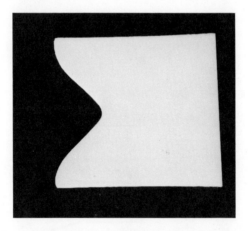

Fig. 7   Polished cross section of experimental double-positive edge contour.

3.3 Blocking Characteristic

The blocking characteristic of a device with a 100-mm dia. wafer is shown in Fig. 8 for various temperatures. The slight asymmetry between forward- and reverse-blocking current can be explained by a different density of recombination centers near the anode and cathode region, respectively. Fig. 9 displays the same characteristic on a logarithmic scale. At lower temperatures ($T_{VJ} \leq 60°C$ in this case), the current initially increases with $\sqrt{V}$, indicating that a thermally generated current density, j_i, in the SCL prevails:

$$j_i = qn_i \frac{W}{\tau_s} \tag{3}$$

where q is the electron charge, n_i the intrinsic carrier density, W the width of the SCL and τ_s the generation lifetime. Quantitative evaluation of τ_s yields typical values between 100 and 400 μs. Obviously, a sharp current increase occurs at higher voltages where injection from the anode junction and carrier multiplication become important.

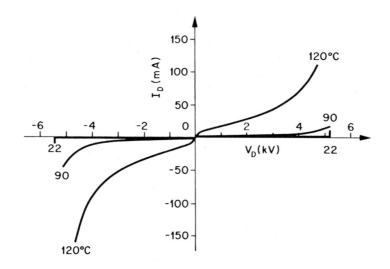

Fig. 8   Forward- and reverse-blocking current at various temperatures on linear coordinates (10-ms pulses).

In most practical applications, high-power thyristors are connected in series. Due to an unavoidable scatter of device parameters, there is a certain time jitter in the turn-off process while reassuming blocking voltage. Those elements turning off comparatively fast may be temporarily overloaded and must be safely protected. The easiest protection, of course, is given if the element itself can take up either an instantaneous overvoltage of as much as 30% above the rated reverse voltage V_{RRM} or a substantial current in the reverse direction. A typical avalanche characteristic for a device rated at V_{RRM} = 4.2 kV is shown in Fig. 10 for various temperatures, demonstrating that such a self-protecting design is quite realistic.

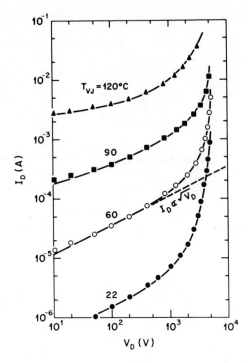

Fig. 9   Forward-blocking current at various temperatures on logarithmic coordinates
          (d.c.).

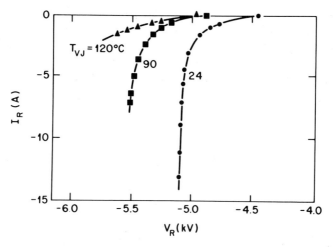

Fig. 10   Typical avalanche characteristic for element with V_{RRM} = 4.2 kV (200-μs pulses).

3.4 Forward Conduction

The current-carrying capability of a power thyristor is intimately connected to the cooling problem, as we have already noted above. Since an efficient cooling system is essential, many high-power applications prefer liquid cooling to air cooling because the thermal resistance of an air-cooled radiator is about twice that of a liquid one.

Although a typical forward-voltage drop is only of the order of 1.5 V, substantial losses occur at each device if we consider a mean average current of 2 kA. To keep these losses to a minimum, the thickness of the silicon wafer has to be small and the carrier lifetime at high current-injection levels should be large. For a blocking voltage of about 4.5 kV, a typical value of the thickness is about 900 μm. As we discuss below, there are limits to this lifetime imposed by the recovery process, more specifically by the recovered charge Q_{rr}, defined in section 4.2, which must lie within narrow specifications. This means that the technology should be adjusted such that a given high-injection lifetime is reached on a reproducible basis; typically 10-20 μs are required.

As an example, data of the average forward-voltage drop of various batches measured at 4 kA and room temperature are plotted in Fig. 11 versus wafer thickness. These values, including total data scatter, are valid for a medium-level Q_{rr} (Q_{rr} = 2.75 mC). A curve for the high-level lifetime has been added which is extrapolated from a theoretical calculation of the forward characteristic.[18] It appears that a lifetime of 16 µs is fairly well approximated in all cases.

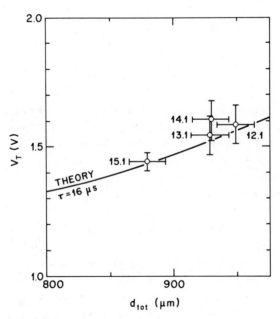

Fig. 11   Average forward-voltage drop, including data scatter of various batches, plotted against wafer thickness. (T_{VJ} = 22°C, I_T = 4 kA, Q_{rr} = 2.75 mC; batch number on graph).

4. DYNAMIC BEHAVIOR

4.1 Turn-On

Parasitic turn-on may be triggered by the displacement current associated with a fast-rising anode voltage, dV/dt. Typically, a dV/dt equivalent to reaching V_{DRM} within one microsecond is required (e.g. 5 kV/µs up to 2/3 V_{DRM} for a 5-kV thyristor). These values can easily be provided by an appropriately shorted emitter. Since a dV/dt firing usually occurs at a small localized spot, the ensuing current (the discharge current of the entire space-charge capacity with an area of order 70 cm^2) destroys the device in most cases. Hence even testing this parameter may be critical.

Although the typical current rise, dI/dt, is moderate (e.g. 100 A/µs), account should be taken of the fact that there is a string of thyristors connected in series which exhibit some scatter in their turn-on times. Hence, the last thyristor to turn on receives a current step which may reach several 100 A with an almost infinite dI/dt as shown in Fig. 12. In order to safely handle currents of this order, a distributed gate

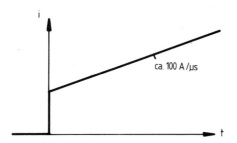

Fig. 12   Initial current as seen by the thyristor that turns on last in a series connection.

is required. Turn-on behavior of an amplifying gate, driven with 400 A/µs, is shown in Fig. 13 where the infrared band-to-band recombination light (λ = 1.14 µm) stroboscopically collected between 0 and 2 µs after voltage breakdown is depicted. The illumination gives a measure of lateral current distribution. The auxiliary thyristor in the center has already passed on its load to the finger structure.

Fig. 13   Gate area of high-power thyristor driven with 400/µs: infrared recombination
light as registered between 0 and 2 µs after voltage breakdown.

4.2 Turn-Off

The turn-off time t_q is limited mainly by the scatter of the turn-off times and the ensuing differences in applied voltage. A typical value for t_q is 0.5 ms, i.e. a small
fraction of a half cycle (t_B = 8.3 or 10 ms). Usually t_q is not a critical parameter,
however. The most important quantity determining the turn-off behavior is the re-
verse-recovered charge, Q_{rr}, already mentioned. In Fig. 14, the definition used in
this context is given for a forward current forced by the external circuit to decrease
with a linear slope (e.g. 3 A/µs). Except for the case where the reverse voltage takes
on very high values, the recovery phase can be theoretically modeled[19] and it has
been shown that the storage time t_{stg} is approximately a constant fraction of the high-
level injection lifetime τ which has been used to model the forward-voltage drop.[20]

If the shaded area in Fig. 14 is approximated by a triangle with a base width proportional to τ, it follows that

$$Q_{rr} \propto \tau^2 \tag{4}$$

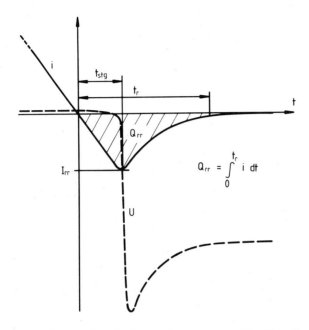

Fig. 14   Definition of the recovered charge Q_{rr} at turn-off with a linearly decreasing
current.

Obviously, equation (4) is valid only if high-level injection prevails throughout the
reverse-recovery phase. During actual turn-off this is not the case and Q_{rr} tends
to depend somewhat more strongly on τ than is predicted by equation (4). By taking
a decrease of lifetime due to the transition to low-level injection into account,[20] it
becomes possible to model the relationship between Q_{rr} and V_T. The result in Fig. 15,
including all the elements of a given batch, shows that the influence of gold diffusion
is fairly well matched. The increase of scattering in Q_{rr} introduced by gold diffusion
is associated only with minor deviations from the theoretical curve. The small scatter
of V_T for a given Q_{rr} may be taken as an indication of a reproducibly homogeneous
current distribution.

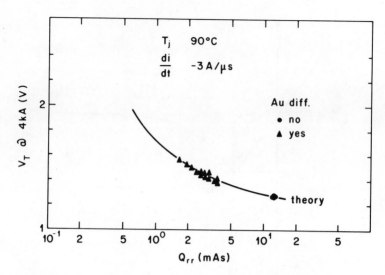

Fig. 15   Forward-voltage drop V_T vs. recovered charge Q_{rr} for an entire batch. The theoretical curve is fitted to the cluster of experimental points without gold diffusion.

It is clear from the explanations above that carrier lifetime is one of the most critical parameters, as has traditionally been the case in the field of power semiconductors. As we have seen, low reverse currents (Fig. 9) impose high generation lifetime ($\tau_s = 100$ to 400 μs) while the turn-off characteristic requires a carefully controlled high-level lifetime ($\tau = 10$ to 20 μs). The problem may be illustrated by the fact that there exists no single, homogeneously distributed recombination center that is known to fit both values simultaneously if Boltzmann statistics are used.[20]

5. FINAL DEVICE

The sheer size of a 100-mm dia. silicon wafer, as shown in Fig. 16, makes packaging more difficult. Since the coefficient of thermal expansion of silicon is much smaller than that of copper (approximately a factor 8), a buffer plate with an intermediate thermal expansion coefficient must be provided (a molybdenum or tungsten plate). Reports about intricate testing of 100-mm dia. wafers alloyed to such a backing plate indicate that this is a critical process.[12] In an alternative approach, these problems may

be avoided altogether by using a symmetrical pressure contact.[21] The so-called concept of "free floating silicon" involves molybdenum discs which are flat to within a few
microns and which are pressed directly onto the semiconductor wafer. Only the evaporated contact metallization (e.g. aluminum) represents a relatively soft layer between
the rigidly pressed partners. In contrast to the alloyed case, an additional thermal
resistance is present which only has influence during short transients and becomes
negligibly small in the steady state (t > 10 ms).[22] Since for high-power thyristors,
only the steady-state value is relevant, a package results with excellent properties
even if it undergoes rigorous thermal cycling. An example of a high-power thyristor
in a casing with 150-mm dia. is shown in Fig. 17.

Fig. 16   Final wafer (100-mm dia.) of a high-power thyristor.

Fig. 17   High-power thyristor ready for use (total size 150-mm dia.; contact area
          100-mm dia.).

6. CONCLUSION

We have seen that blocking voltages of high-power thyristors have generally reached a level between 4 and 5 kV. The typical mean-average current of such thyristors manufactured from 100-mm dia. wafers is about 2 kA (80°C case temperature). The higher values cited in some publications invariably result from arbitrarily assumed lower case temperatures.

We conclude by inferring possible future trends on the basis of the previous discussion of the various device parameters:

- Lifetime is a very important quantity which controls the tradeoff between voltage drop, V_T, and recovered charge, Q_{rr}. This tradeoff is very similar to that required for adjusting the turn-off time, t_q, of a fast-switching thyristor. Hence the problem of optimizing a high-power thyristor is very similar to that of a frequency thyristor operating at substantially higher voltages. This optimum is carefully balanced from a systems point of view. Hence, any change of such external parameters as the blocking voltage will lead to a new optimization.

- From the standpoint of most applications, a further increase in blocking voltage is desirable. This means, however, that either the static forward losses or the size of the recovered charge, Q_{rr}, must be allowed to increase. As a result there will be practical limits, quite apart from those on device behavior, of how much increase in thyristor voltage will be of advantage.

- An alternative future device design might involve asymmetric structures such as are currently used for reverse-conducting thyristors.[23] Potentially, there is the possibility of increasing the blocking voltage by almost a factor of two without impairing forward losses and turn-off properties. The problem will be to change the system and/or the circuit such that is allows for operation with no reverse voltage applied to the thyristor.

- The current-carrying capability with a 100-mm dia. wafer is large enough for HVDCT-power conditioning, up to approximately 1000 MW, without paralleling thyristors. Since it is assumed that the total power handled will not increase dramatically in the next five years, it may be expected that the switching power, P_s, (see Fig. 3) will grow at a slower pace than hitherto.

- As more and more emphasis is placed upon conservation of energy, the importance of high-power thyristors will certainly increase.

REFERENCES

1. M. Bosch et al., "Ergebnisse der Prüfung von HGUe-Ventilen und Einsatz der Forschungsanlagen in Rheinau," CIGRE Int. Conf. on Large High Tension Electric Systems, 10-20 June 1968, Paper 43-05.

2. I. Boban et al., "Design and Testing of the Cabora Bassa Thyristor Valves," CIGRE Int. Conf. on Large High Voltage Electric Systems, 21-29 August 1974, Paper 14-07.

3. A. Herlet, "Hochleistungsthyristoren," Elektro-Anzeiger, 32, No. 7 (1979) 48-53.

4. O.M. Clark, "Voltage Breakdown of Silicon Rectifiers as Influenced by Surface Angle," J. of the Electrochem. Soc., 107 (December 1960) 269 C.

5. R.L. Davies and F.E. Gentry, "Control of Electric Fields at the Surface of p-n Junctions," IEEE Trans. Electron Devices, ED-11, No. 7 (1964) 313-323.

6. M. Schnöller, "Breakdown Behavior of Rectifiers and Thyristors Made from Striation-Free Silicon," IEEE Trans. Electron Devices, ED-21, No. 5 (May 1974) 313-314.

7. J. Cornu, S. Schweitzer, and O. Kuhn, "Double Positive Beveling: A Better Edge Contour for High-Voltage Devices," IEEE Trans. Electron Devices, ED-21, No. 3 (March 1974) 181-184.

8. M. Bakowski and B. Hansson, "Influence of Bevel Angle and Surface Charge on the Breakdown Voltage of Negatively Beveled Diffused p-n Junctions," Solid State Electronics, 18 (1975) 651-657.

9. R.O. Evans, "Microelectronics in Data Processing and Telecommunications," from Electronics to Microelectronics, Eurocon 80, Stuttgart, March 24-28, 1980, 18-28.

10. K. Brisby, "Thyristors for HVDC Applications," ASEA J., 52, No. 3 (1979) 68-71.

11. M. Morita et al., "Large-Area, High-Voltage Thyristor for HVDC Converter," Internat. Electron Devices Meeting, Washington D.C., December 1977, 26-29.

12. K. Kamahara et al., "4000 V, 2500 A High-Voltage High-Power Thyristor," Industrial Applic. Soc. (1978) 1022-1028.

13. S.M. Sze, Physics of Semiconductor Devices, New York: J. Wiley, 1969.

14. T.F. Ciszek, "Solid-Liquid Interface Morphology of Float Zoned Silicon Crystals," Semiconductor Silicon, New York: The Electrochem. Soc., 1969, 156-168.

15. M.J. Hill, P.M. van Iseghem, and W. Zimmermann, "Preparation and Application of Neutron Transmutation Doped Silicon for Power Device Research," IEEE Trans. Electron Devices, ED-23, No. 8 (August 1976) 809-813.

16. J.M. Meese, "The NTD Process - A New Reactor Technology," in Neutron Transmutation Doping in Semiconductors, New York: Plenum Press, 1979, 1-10.

17. J. Cornu, "Field Distribution Near the Surface of Beveled p-n Junctions in High-Voltage Devices," IEEE Trans. Electron Devices, ED-20, No. 4 (April 1973) 347-352.

18. J. Cornu and M. Lietz, "Numerical Investigation of the Thyristor Forward Characteristic," IEEE Trans. Electron Devices, ED-19, No. 8 (August 1972) 975-981.

19. C. Abbas, M. Lietz, and R. Sittig, "Turn-off Behavior of Power Recitifiers," from Electronics to Microelectronics, Eurocon 80, Stuttgart, March 24-28, 1980, 364-366.

20. J. Cornu, R. Sittig, and W. Zimmermann, "Analysis and Measurement of Carrier Lifetimes in the Various Operating Modes of Power Devices," Solid State Electronics, 17 (1974) 1099-1106.

21. S.D. Prough and J. Knobloch, "Solderless Construction of Large Diameter Silicon Power Devices," IEEE Industrial Applic. Soc. Meeting, Los Angeles, October 2-6, 1977, 817-821.

22. M.S. Adler, "Fast Power Diodes," present volume.

23. D. De Bruyne, J. Vitins, and R. Sittig, "Reverse-Conducting Thyristors", present volume.

DISCUSSION
(Chairman: Dr. W. Gerlach, TU, Berlin)

W. Gerlach

If the lifetime is not homogeneous in large-area devices then you might have differences in plasma spreading in various directions. Is this a problem in the manufacturing of these devices?

A. Jaecklin (BBC Brown, Boveri & Co. Ltd., Baden)

I should first say that there is inhomogeneity in two directions. First, there is an axial inhomogeneity which we actually want; secondly, there is the lateral inhomogeneity which I think is what you are referring to. Measurements have been made to clarify the extent of these problems. The best way of coping with them is to determine the magnitude of the additional losses introduced by plasma spreading. For a half wave with a 4-kA peak and 10 ms basewidth, we have measured additional losses of less than 10 %. We found a significant dependence on the reverse-recovery charge which is connected with the lifetime as you would expect, but it is a problem we have solved.

D. Silber (AEG-Telefunken, Frankfurt)

Can you comment on actual lifetime gradient?

A. Jaecklin

No, I wouldn't like to comment on that.

H. Irmler (Semikron GmbH, Nürnberg)

I recall some old work of Hoffmann at Siemens which indicated a linear relationship between Q_{RR} and τ, whereas you said Q_{RR} is proportional to τ^2.

A. Jaecklin

This is actually very easy to interpret. Essentially the time delay between the zero crossing of the current and the starting of the voltage that appears is proportional to the lifetime. That has been correlated very well by numerical analysis made here at the BBC Research Center. Now Q_{RR} is the area of that shape, which can be approximated by a triangle and since you have a linear slope at the beginning, you automatically get τ^2 into your equation.

H. Melchior (ETH, Zurich)

Is it correct to assume that the lifetime of the high-injection levels is more or less independent of the current or the carrier density?

A. Jaecklin

In a publication by Cornu, Sittig, and Zimmermann[1] several years ago the injection dependency of lifetime was measured for power devices. They found that at a certain injection level you get a saturation and essentially you can approximate this high-level injection lifetime by the addition of τ_{no} and τ_{po} according to Shockley-Read-Hall theory. You have a completely different relation if you take the generation lifetime where you get amplification factors which are very great, but I think that goes beyond this discussion.

J. Gobrecht (Brown Boveri Research Center, Baden)

In one of your first slides you showed that more than 90 % of the fast neutrons bring silicon atoms to interstitial sites. But after this process the neutrons have sufficiently high speed to transmute silicon to phosphorus and therefore the balance you have shown is incorrect. Must it not well exceed 100 %?

A. Jaecklin

You have to interpret that rather carefully. Although there are a certain number of fast neutrons in every reactor, the pool reactor is the one that has the lowest percentage. It's just the damage by those fast neutrons which do not really lead to transmutation doping, but essentially the slow ones that lead to it. The figure I have given is just the number of displaced silicon atoms and there are measurements and computer simulations of this where you create small areas of essentially amorphous silicon and it's just all this damage calculated together. The figures I have given are based on theoretical evaluations.

J. Slatter (Philips Research Laboratories, Redhill)

What do you think to be the optimum distribution in lifetime along the axis?

A. Jaecklin

This again is a question I have not worked out in too much detail but what you want is a long lifetime both in blocking and in forward conduction, and a short lifetime for reverse recovery. This means that reverse recovery is concerned with the anode-side junction, whilst blocking and forward drop are essentially concerned with the cathode-side junction. Thus you would like to have different levels there.

REFERENCES

1.   J. Cornu, R. Sittig, and W. Zimmermann, "Analysis and Measurement of Carrier Lifetimes in the Various Operating Modes of Power Devices," Solid State Electronics, 17 (1974) 1099-1106.

LIGHT-ACTIVATED THYRISTORS[*]

D. SILBER, H. MAEDER, M. FUELLMANN
AEG-Telefunken, Frankfurt, Fed. Rep. Germany

SUMMARY

Perfect isolation between power and control circuits in high-voltage converter equipment is obtained by optical thyristor triggering. Direct optical triggering of power
thyristors and auxiliary thyristors with lasers or LEDs, i.e. triggering with an
internally generated photocurrent, requires the development of highly trigger-sensitive
thyristor gates. High trigger-power amplification and different methods of avoiding or
compensating for fault triggering permit the design of light-activated thyristors which
can be triggered with light power in the milliwatt range. The temperature dependence
of turn-on properties and some relations to thyristor structures can be obtained from
relatively simple charge-control model calculations.

1. INTRODUCTION

Optical triggering of power thyristors is a very attractive technique for high-voltage
and noise isolation between the power and trigger circuits. Basically there are two
ways of optical triggering. The first is "indirect" optical triggering, which means
optical transmission of the trigger signal, but not the trigger power. Some additional
circuitry for trigger-power transmission and switching on the thyristor cathode potential is necessary in this case. Circuits of this type are well known in both high-voltage (HVDC transmission) and low-voltage (solid state relay) applications. The second
approach, which is the subject of this paper, is direct optical triggering, which means
optical trigger-power transmission, and integration of an optically sensitive gate into
the thyristor. Another, in some sense intermediate, possibility is auxiliary thyristor
triggering. In this case, a small optically triggerable thyristor is coupled to a conventional power thyristor as an external gate amplifier. This permits the use of a few
types of optically triggered devices combined with a wide variety of power thyristors
and allows the addition of a protecting network, e.g. breakover diodes (see Fig. 1).

---

[*] Presented at the symposium by D. Silber

(a)

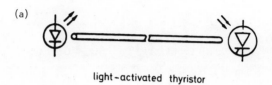

light-activated thyristor

(b)

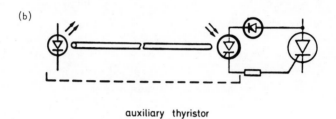

auxiliary thyristor

(c)

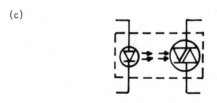

bidirectional thyristor (compact)

Fig. 1   Thyristor triggering by internally generated photocurrent.

Whereas optically triggered low-voltage thyristors have been known for a long time, optical triggering of high-voltage thyristors was rarely discussed in the literature prior to 1970. The investigations and developments were motivated by the development of very high-voltage circuits, especially HVDC transmission and V-A reactive genera-tors.[1] The first papers dealing with the special properties and design of optically triggered high-voltage thyristors appeared in 1975[2] and 1976,[3-6] and there were also important contributions earlier in the patent literature.[7]

In this paper we describe the present status of development and give a short treatment of both the basic problems which had to be solved and the basic ideas which have led to successful development of new gate concepts.

2. LIGHT SOURCES AND LIGHT-TRANSMISSION SYSTEMS

In direct optical triggering of thyristors, irradiated light generates a photocurrent in the high-voltage device. The highest conversion efficiency is obtained when the carriers are generated in the space-charge layer and in the p-base because of their drift fields. As a consequence, light with a penetration depth of several tens of microns in silicon, corresponding to wavelengths between 850 and 950 nm, is most appropriate.[8] Until now, GaAs light-emitting diodes or lasers have been the light sources considered for most applications. For optimum choice of the light source, the following points have to be considered:

a) The optical power density irradiated within a limited aperture must be sufficiently high to permit efficient light transmission.

b) The long-term stability of the light source must be proven.

TABLE 1

Properties of Different Light Sources Used for
Direct Thyristor Triggering

Light Source	Advantage	Disadvantage
GaAs pulse lasers	high pulse power; convenient coupling even into single fibers	pulses; high drive current pulses
GaAs-GaAlAs DHS lasers	continuous power; convenient coupling even into single fibers	expensive; sensitive
GaAs (:Si)-LED heterostructure power LED	cheap	low rad. density; coupling problems

In Table 1, the basic properties of different light sources are compared. Although the CW laser would obviously be an ideal light source, its cost and critical operating conditions have so far prohibited its application for thyristor triggering.

The basic problem of the conventional GaAs power LED is efficient coupling to glass-fiber cables or glass rods. Fig. 2 shows the calculated coupling loss and the transmission loss for a light source with Lambert-type radiation characteristics coupled into glass cables with different attenuation and numerical aperture (num.ap.). The space factors and reflection losses have been neglected. We have considered a special large-aperture cable, a cable for short-range signal transmission, and a low attenuation cable. Oviously, for short transmission distances, for example for photocoupler-type combinations of light source and thyristor, a high aperture is the essential requirement. Glass rods which make use of total reflection at the glass-air-boundary are perfect transmission systems in this respect. For glass-cable lengths above 5 meters, low attenuation becomes the major requirement.

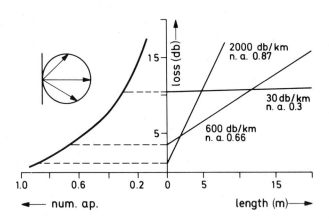

Fig. 2   Coupling and attenuation losses in different glass-fiber cables coupled to a Lambert-type radiator.

In practice, power LEDs do not have exact Lambert-type characteristics. The very powerful hemispherical GaAs LEDs especially have a nearly constant angular distribution which is even more disadvantageous in this respect. Lasers, however, can be coupled even to a low-aperture cable with very low coupling loss. Furthermore, the high power density of the laser enables coupling to single glass fibers a few hundred microns in diameter, whereas LEDs have to be coupled into glass-fiber bundles or thick-core fiber cables. This is the reason for the superior properties of laser systems which will become more impressive when the advantage of high trigger-power density at the thyristor is considered.

An interesting way of obtaining good transmission of LED radiation over a distance of several meters is the use of glass cones for trigger light concentration[9,10] (Fig. 3). From fundamental optical laws it follows that the radiation aperture is correspondingly enhanced and therefore only radiation transmitted with a low aperture can be concentrated. By using a cone at the LED, high-aperture radiation can also be converted to a lower aperture. The light must then be transmitted with relatively thick glass cables (2 - 3 mm diameter) and can be reconcentrated at the device.

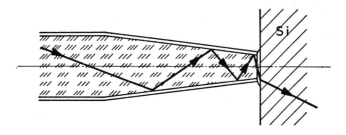

Fig. 3 Concentration of light with glass cones.

For optimization of very high-aperture light-transmission systems, the coupling materials between the glass and semiconductor surfaces are important. They not only reduce reflection from the silicon surface, but can also enhance the light power coupled out from the glass rod or glass cone because of the prevention of total reflection of high-aperture radiation at the end surface.

The light power which can be obtained from Si-doped GaAs LEDs coupled into 1 mm-diameter high-aperture glass cables or glass rods can be in the range of a few milliwatts CW power, and several tens of milliwatts for pulses with 100 μs pulse lengths at line-frequency repetition rates. To ensure long-term stability, however, it may be necessary to reduce the LED drive current well below the recommended data-sheet values. The attenuation losses, and the reduced coupling efficiency to low attenuation glass cables, must also be considered.

GaAs pulse lasers can provide a pulse energy of several hundred nanojoules with conventional glass cables.[6] Available GaAs-GaAlAs CW lasers would yield about 5 mW continuous optical trigger power to the thyristor.

3. SPECIAL PROBLEMS OF OPTICAL THYRISTOR TRIGGERING

As shown in the previous section, photocurrent triggering of thyristors corresponds to triggering with very low power and, in the case of LEDs, also to low power density. As a consequence, the thyristors have to be designed for efficient photocurrent generation, high internal trigger-current amplification to yield:

a) Sufficiently short turn-on delay and, as far as possible, independence to moderate light-power variations, which may arise from LED degradation or from light cable coupling properties.

b) Reasonable dI/dt capability, including the capability of snubber-circuit discharge at high voltage.

Such high trigger sensitivity in conventional thyristors would, however, result in correspondingly high sensitivity to fault triggering due to internally generated capacitive currents. The main difficulties would be:

c) dV/dt loads (capacitive currents)

d) "rest plasma enhanced" capacitive currents arising in commutating dV/dt in triac structures (or the recurrence of the forward-blocking voltage after thyristor turn-off).

Some important basic relations can be obtained from a simple optically sensitive amplifying gate structure (Fig. 4). Trigger light is irradiated through the cathode n^+ emitter in this example (this is not the case in all published gate concepts; some authors[4,6] have preferred direct irradiation into the p-base or into the space-charge region[3]). Efficient photocurrent generation in the base regions is obtained when the absorption in the n^+ emitter is considerably reduced by reduction of the zone thickness to about 6 µm.[5] This point has been more systematically investigated by Konishi et al.[11] The corresponding deterioration of the emitter properties (emitter h parameter[12]) would not affect the trigger sensitivity because it is only effective at high current density, and the increased emitter-sheet resistance increases the transverse field-emitter voltage thereby improving the dI/dt capability (see section 7).

Carrier-pair generation in the p-base, the space-charge region and the n-base, all result in trigger currents being the same for both bases, and therefore shows great similarity to the capacitive fault trigger current caused by dV/dt load. A simple but important result is obtained from a calculation of the trigger potential shape in the p-base caused by either an optical or a dV/dt-trigger current.[5] The potential profile is shaped parabolically in regions with carrier generation, and logarithmically outside. The n^+ emitter potential is shifted to the p-base potential height at the emitter shunts. A usual condition of optical triggering is light irradiation with constant power

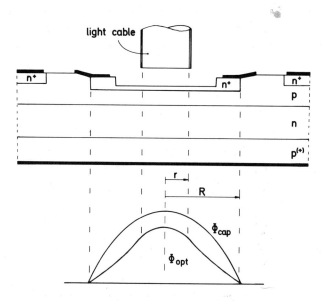

Fig. 4   Optically generated (ϕ_{opt}) p-base trigger potential and dV/dt load capacitive p-base potential (ϕ_{cap}) in a simple optically triggered thyristor gate structure.

density into a fixed area, for example from a light-fiber cable with $2\,r = 1$ mm diameter. If we look for the possibility of obtaining high trigger sensitivity, we can increase the p-base sheet resistance which leads to a proportional increase of the trigger potential, or we can increase the emitter shunt radius R which results in an increase of the logarithmic part of the trigger potential.

If we add the further condition of minimizing the fault-trigger potential, we obtain the obvious result that an optimum is obtained when $r \geq R$. Any deviation from this leads to a "quality reduction factor" f_Q

$$f_Q = (r/R)^2 \cdot \{ 1 - 2 \ln (r/R) \}$$

The result, that the "fault trigger-current generation" area should not be larger than the "trigger-current generation area" in the light-activated thyristor gate holds for most optically triggerable gate structures, and when fulfilled, results in surprisingly high dV/dt capabilities and good optical trigger sensitivity of high-voltage thyristors (Fig. 5). In the following discussion of the relations between optical trigger sensitivity and dV/dt fault-trigger sensitivity, we assume that this optimization condition is fulfilled.

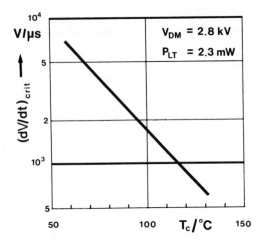

Fig. 5   dV/dt capability vs. temperature obtained in a thyristor gate according to Fig. 4 with R $\approx$ r = 0.5 mm.

For thyristors with small shunting radius R and very high trigger sensitivity there is, however, a fundamental limitation to this circular structure. The total amount of holes in the p-base per unit area is small because of the necessarily high sheet resistance and, at high voltage, in a similar way to a junction FET, it is reduced considerably by extension of the space-charge layer. For p-base sheet resistances above 10^5 $\Omega/\square$, this results in p-base punchthrough.

To remain close to the optimization condition in small structures with high trigger sensitivity, the circular symmetry must be discarded. Masked p-base diffusions to obtain high-resistance lateral arms are a possible solution. Structures of this type have been proposed by De Bruyne and Sittig,[3] Jaecklin,[13] and Temple.[14]

4. A SIMPLIFIED MODEL FOR TURN-ON BEHAVIOR

Detailed modelling of the turn-on behavior of optically sensitive devices has been made by Temple[15] using charge-control equations. We present a slightly different and simplified version of this model, but include the influence of emitter shunting. Attention will be concentrated on a most important factor which is often neglected, that is the large temperature dependence of the trigger properties.

The charge-control model is illustrated in Fig. 6. Although it is possible to include the emitter charges Q_1^* and Q_2^* in analytical solutions, for present purposes the simplified version is sufficient.

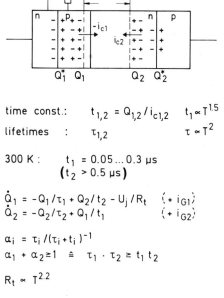

time const.: $\quad t_{1,2} = Q_{1,2} / i_{c1,2} \quad\quad t_1 \propto T^{1.5}$

lifetimes : $\quad\quad \tau_{1,2} \quad\quad\quad\quad \tau \propto T^2$

300 K : $\quad t_1 = 0.05 \ldots 0.3 \ \mu s$
$\quad\quad\quad (t_2 > 0.5 \ \mu s)$

$\dot{Q}_1 = -Q_1/\tau_1 + Q_2/t_2 - U_j/R_t \quad (+ i_{G1})$
$\dot{Q}_2 = -Q_2/\tau_2 + Q_1/t_1 \quad\quad\quad (+ i_{G2})$

$\alpha_i = \tau_i /(\tau_i + t_i)^{-1}$
$\alpha_1 + \alpha_2 \gtrless 1 \quad \hat{=} \quad \tau_1 \cdot \tau_2 \gtrless t_1 t_2$

$R_t \propto T^{2.2}$

Fig. 6 Illustration of the charge-control model (Q_1^* and Q_2^* have been neglected in this calculation).

The charge-control time constant can be calculated from standard minority-carrier transport equations in the p-base, and is strongly dependent on the doping profile. Assuming an exponentially decreasing p-base diffusion profile, $N_A \sim \exp(x/l_p)$, and p-base width w_p, the time constant is

$$t_1 = l_p^2/D_n \cdot \{ \exp(-w_p/l_p) - (1 - w_p/l_p) \}$$

where the reduction of w_p from space-charge regions is neglected.

Typical t_1 values at room temperature for n^--doping of $3 \cdot 10^{13}$ cm^{-3} and a p-base sheet resistance 2000 $\Omega/\square$ are 300 ns for $w_p = 85$ μm and 50 ns for $w_p = 38$ μm.

The time constant t_2 is difficult to define in thyristor charge-control models because of drift components and extension of the space-charge layer in the n-base. We assume it to be few microseconds (at least 1 μs) at high voltage, whilst at low voltage it will become much larger (section 5). Fig. 6 shows the assumed temperature dependences of t_1 and of both minority-carrier lifetimes.

The emitter shunt has been treated in a simplified but reasonable approximation by assuming constant (but strongly temperature-dependent) shunt resistance. The Boltzmann voltage is assumed to be 0.5 V at 300 K and 0.35 V at 400 K. The p-base sheet resistance temperature dependence is assumed to be proportional to $T^{2.2}$ because of the low p-base doping concentration.

When discussing the trigger limits, we have to distinguish two important approximations

    a) Photocurrent triggering
        In this case the gate currents I_{G1} and I_{G2} are equal. This approximation may also be useful for the discussion of dV/dt triggering of thyristors with low dV/dt (rise time long compared to $\sqrt{t_1 t_2}$).

    b) Charge triggering
        This is an approximation for optical pulse laser triggering, where the p-base trigger charge Q_1 is larger than Q_2, and for the discussion of thyristor triggering with high dV/dt, where $Q_1 = Q_2$ (neglecting the emitter charge).

The most important results are given in Table 2.

The photocurrent trigger sensitivity is strongly temperature dependent mainly because of the high temperature dependence of the p-base sheet resistance. In the numerical example, $R_t = 2 \Omega$, (normalized to 1 cm^2 area). The p-n-p amplification component τ_2/t_2 has been assumed to be about 0.25, which is reasonable for triggering high-voltage thyristors under low applied voltage (and therefore large t_2).

TABLE 2

Results of a Simplified Charge-Control Calculation

(R_t is assumed to be 2 Ω because of normalization

to 1 cm^2 carrier-generation area)

Photocurrent triggering, low dV/dt

$$I_G \underset{\sim}{\sim} U_j/R_t \ (1 + \tau_2/t_2)^{-1}$$

0.2 A/cm^2 (300 K)

0.07 A/cm^2 (400 K)

Charge triggering, high dV/dt

(long lifetime approx.)

$$Q_1/\sqrt{t_1 t_2} + Q_2/t_2 \underset{\sim}{\sim} U_j/R_t$$

for $Q_1 \underset{\sim}{\sim} Q_2$

$(0.4...1.7) \ 10^{-7}$ Cb (300 K)

$(0.2...0.8) \ 10^{-7}$ Cb (400 K)

dV/dt charge: 10^{-7} Cb

Trigger delay time

(long lifetime approx.)

$$t_d \propto \sqrt{t_1 \ t_2}$$

$\sqrt{t_1 t_2} \underset{\sim}{\sim} 0.2...1.0$ μs

(at high voltage)

The example for charge triggering has been calculated for the same R_t, but for devices with different time constants t_1. The temperature dependence is almost as strong as in the case of photocurrent triggering. A further interesting result is that for "slow" devices, even at 400 K, the critical trigger charge is approximately 10^{-7} Cb, which is almost the total capacitive charge (about $1.5 \cdot 10^{-7}$ Cb at a maximum field strength of $1.5 \cdot 10^5$ V/cm). The dV/dt capability can thus be extremely high. It should be noted, however, that the device is shunted to 2 mA/mm^2 photocurrent trigger level which is not especially low ($\hat{=}$ 4 mW/mm^2 trigger power).

It is concluded that attempts to obtain exact dV/dt trigger levels from trigger-current calculations are far from realistic for "slow" devices with high dV/dt capability. The disadvantage of the "slow" device, however, is its extended turn-on delay time, which may be assumed to be approximately proportional to the cycling time $\sqrt{t_1 t_2}$.

Our results illustrate a further important fact. High dV/dt capabilities can be achieved much more easily in high-voltage than in low-voltage thyristors at the same optical trigger sensitivity. Because of the slightly reduced maximum field strength of the high-voltage devices (which results in reduced capacitive charge) and also the larger time constants, the dV/dt capability can be enhanced more than proportionally to the device blocking voltage.

5. EXPERIMENTAL RESULTS CONCERNING THYRISTOR TRIGGER PROPERTIES

In the previous section, the influence of p-n-p photocurrent amplification on the trigger sensitivity was mentioned. Fig. 7 shows photocurrent response measurements taken from two devices (thyristors with removed n[+] emitters). We have also included the photoresponse of corresponding p-n diodes (thyristor without n[+] and p-emitter). The

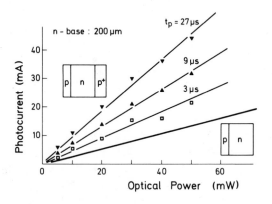

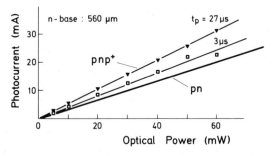

Fig. 7   Photocurrent amplification in two different p-n-p[+] structures.

photocurrent amplification has a large time delay in the high-voltage device. It is concluded that for triggering within a few microseconds, the p-n-p photocurrent amplification can be neglected in high-voltage devices, but not in thyristors with 1200-V blocking voltage.

Surprisingly good agreement is obtained when, according to the discussion in the previous sections, the observed temperature dependence of the trigger level is compared with the well-known carrier mobility variation and with a simplified expression for the Boltzmann junction voltage at the trigger level (Fig. 8). We have also found a slight increase in the trigger delay time at increased temperatures. Fig. 9 shows experimental results obtained from high-voltage auxiliary thyristors. It should be noted, however, that for reasons which are not understood, this is not in agreement with results obtained by Jaecklin.[13]

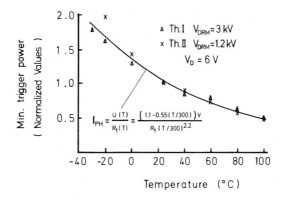

Fig. 8   Temperature dependence of thyristor trigger level.

Fig. 5 shows the variation of the critical dV/dt with temperature. This behavior cannot be directly compared with calculated results because the total capacitive charge is held constant and only the rise time is varied.

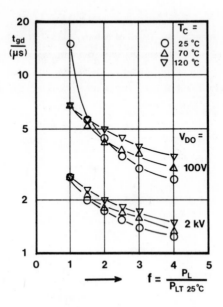

Fig. 9   Trigger delay time vs. overdrive factor for a 3-kV light-activated thyristor.

6. SPECIAL GATE CONCEPTS FOR HIGH dV/dt CAPABILITY

In the previous sections we have discussed general properties of light-triggered thy-ristors. Optimization of the relation between trigger-power sensitivity and dV/dt ca-pability, masked p-base diffusions for high lateral resistances, and much reduced n^+ emitter thickness are special thyristor-gate concepts which have not been used previously with respect to power thyristors. The dV/dt fault-trigger problem and the demands for high dI/dt capability have led to additional fundamentally new concepts of gate structures.

6.1 Special Gate Structures with Reduced dV/dt Trigger Sensitivity

The discussion of amplifying gate optimization in section 3 assumed conventional thyristor structures. Reduced p-base capacitance can be obtained when the p-base is formed by masked grid diffusions (see De Bruyne and Sittig,[3] Fig. 10).

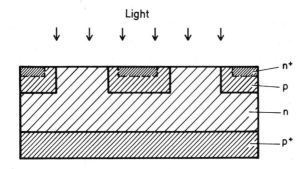

Fig. 10   p-base grid light triggerable gate structure.

As the space-charge region at the grid reaches the surface, very high conversion efficiency is obtained even for light with a shorter wavelength and low penetration depth. The curvature of the space-charge boundary results in a 10 % decrease of the forward blocking voltage. Although this seems to be a disadvantage, it is just this which permits breakover trigger capability, because the avalanche current is amplified in the same way as the optically generated trigger current.

The p-base photocurrent is focused to a small integrated thyristor, so that the total concept is an amplifying gate thyristor with an integrated p-grid photodetector. Further thyristor concepts based on this idea have been reported by De Bruyne et al.[16] and Jaecklin and Bajan.[13] In high-voltage power thyristors, trigger power levels P_{LT} of about 2 mW could thus be obtained with dV/dt capabilities of about 2 kV/μs. Auxiliary thyristors (V_{DRM} = 4 kV) with a maximum junction temperature of 70°C have been obtained with $P_{LT} \cong$ 2.5 mW and dV/dt $\cong$ 5 kV/μs.[16,13]

A further development for reduced p-n junction capacitance is the inclusion of inhomogeneous n-base doping.[16,17] Thyristors with p-i-n (or rather p-n⁻⁻-n) base structures exhibit reasonably higher total voltage in structures with the same total base thickness.[1,18] For junction-capacitance reduction, a reduced maximum electrical-field strength can easily be obtained in such structures, so that the total capacitive charge and the corresponding sensitivity to dV/dt fault triggering are reduced (Fig. 11).

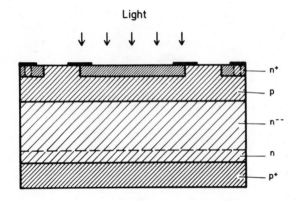

Fig. 11   p-n⁻⁻-n base type thyristor structure.

This combination of p-base grid structures with p-i-n type base doping is an obvious concept because the large extent of the space-charge regions should reduce the breakover voltage diminution. Considering that the p-i-n thyristors do not exhibit reverse-blocking voltage, their use in high-voltage applications may be restricted to auxiliary thyristors.

6.2 Gate Structure with Switched Trigger Sensitivity

In section 3, we discussed the dependence of thyristor trigger sensitivity on the emitter-shunt resistance. Oviously, high emitter-shunt resistance would be useful when the device is to be triggered optically, and low resistance in the case of dV/dt loads. A straightforward way of obtaining such behavior is emitter-shunt switching (Fig. 12).

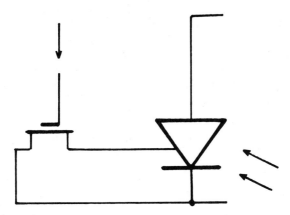

Fig. 12   MOS-transistor switched emitter shunt.

The use of integrated lateral MOS transistors to achieve such behavior has been proposed by Svedberg,[19] O'Neill et al.[20] and more recently by Patalong et al.[21] The switching-gate voltage would have to be obtained, for example, from dV/dt capacitive currents. O'Neill's device is an optically triggered, low-voltage, low-current triac driver. In this case, the total applied voltage is transferred to the MOS transistor gate. Thus the trigger sensitivity is largely reduced at device voltages above 20 V which correspondingly reduces both "static" dV/dt and "commuting" dV/dt trigger sensitivity. The inability to trigger the device optically above this voltage is an advantage for devices which are used for triggering at zero-crossing voltage only.

For devices used in high-voltage applications, however, it has to be ensured that optical triggering is possible under all voltage and dV/dt conditions. This is obvious if, for example, series strings are considered. A strong reduction in optical trigger sensitivity under dV/dt conditions would further enhance the trigger delay time of the "slowest" device in the string. Therefore, the amount of trigger-sensitivity reduction and the emitter-shunt switching conditions have to be designed very carefully. The main advantage of the MOS-switched emitter shunt may be the possibility of integrating it into very small areas.

6.3 dV/dt Compensation Concepts[2,22,23]

In comparison to the reduction of trigger sensitivity in the optically sensitive parts of the device under dV/dt conditions, an alternative would be the integration of some gate-turn-off system which exactly compensates the fault-trigger current or pulse charge by extracting the same current or charge from the trigger base. In this way, fault triggering could be prevented without any reduction of the optical-trigger sensitivity. A simple circuit exhibiting this property is shown in Fig. 13a. The compensation condition is

$$R_c \, C_c = R_i \, C_i$$

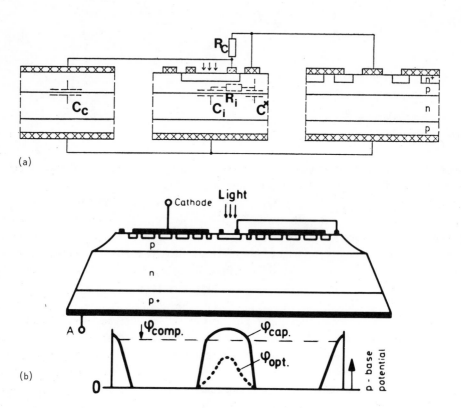

(a)

(b)

Fig. 13   Equivalent circuit and schematic view of a dV/dt-compensated light-activated thyristor.

i.e., the base and the n^+ emitter potentials are equal, and forward biasing of the n^+ emitter junction is prevented. Integration into power thyristors is straightforward, and Fig. 13 b shows the design of an optically triggered power thyristor according to this principle. Its operation is demonstrated by discussing the gate-trigger potentials. In conventional amplifying gate-type integrated pilot thyristors, the n^+ emitter potential with respect to the trigger base potential is determined from the location of the emitter shunt (Fig. 4). According to the discussion presented in section 3 it is then possible to obtain some optimum compromise between optical and dV/dt trigger sensitivity, but total insensitivity to fault triggering could only be achieved by shunting the n^+ emitter at the point of maximum fault-trigger potential. This, however, would make the device correspondingly insensitive to optical triggering. In the dV/dt compensated device shown in Fig. 13b, the n^+ emitter is shunted to the p-base in a different area (the edge area in this example), which under dV/dt load exhibits the same potential enhancement. Thus dV/dt triggering is prevented, but in the case of optical irradiation, there is no influence on the p-base in this "compensation" area and forward biasing of the n^+p emitter junction is not prevented.

If, however, R_c approaches the magnitude of R_i, the turn-on properties deteriorate because a strong negative feedback during turn-on is produced. Therefore, C_c must be several times larger than C_i, and this determines the minimum integration area for this type of dV/dt compensation.

In small-area devices, such as auxiliary thyristors, the compensation may be performed by "asymmetric" emitter shunting (Fig. 14). The thyristor gate is designed in such a way as to produce additional high p-base fault-trigger potential in an area which is different from that of the trigger potential maximum. Emitter shunting in this first area almost entirely prevents dV/dt fault triggering but has little influence on optical-trigger sensitivity. An advantage of integrated dV/dt compensation is the matched thermal compensation of R_c and R_i which is associated with it.

A further development of the compensation principle includes a p-base potential-limiting diode as shown in Fig. 15a. In this case, the optical-trigger sensitivity is approximately determined by the sum of the resistors R and R*. To avoid fault triggering, it is sufficient to keep the n^+p emitter forward bias below the threshold voltage of the diode. The compensation condition

$$R_c\, C_c > R_i\, C_i$$

is then independent of R* and also of trigger sensitivity. Fig. 15b shows a compensation performed with a hybrid potential-limiting diode.[22]

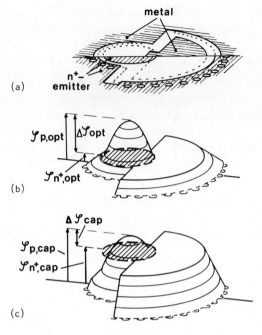

Fig. 14   dV/dt compensation by asymmetrical emitter shunting:
           a) gate structure
           b) optically generated p-base and n$^+$ emitter potentials
           c) capacitively generated potentials during dV/dt load

Further advantages of this concept are:

a) R_c < < (R_i + R*) can be easily achieved with relatively low R_i and a corres-
    pondingly low compensation area for C_c.

b) The integration of high R* for high trigger sensitivity can be performed with-
    out respect to the "balasting" capacitance C*. This is useful for the design of
    gate structures with very high optical sensitivity, because complicated masked
    p-diffusions to achieve high lateral resistances in small areas can be avoided
    or simplified.

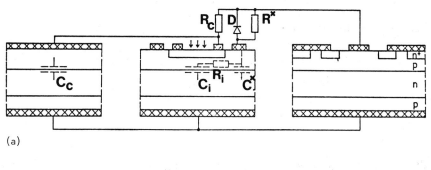

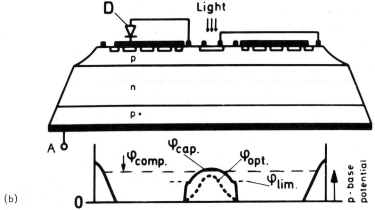

Fig. 15   dV/dt compensation including p-base potential limiting diode.

Light-activated thyristor gates of this type have been designed for $P_{LT} < 1.5$ mW in a gate area of 1 mm^2 and with a dV/dt capability > 1 kV/μs at 125°C.

7. SPECIAL CONSIDERATIONS FOR HIGH INTERNAL TRIGGER-CURRENT AMPLIFICATION

The high dI/dt capability of conventional electrically triggered thyristors requires a high trigger current, especially for extended gate structures. To overcome this, various gate concepts with internal trigger-power amplification were proposed between 1965 and 1970 which can be basically classified into amplifying gate and field-emitter type structures.[24]

In optically triggered thyristors, very effective trigger-power amplification is necessary because of the low trigger photocurrent. In addition, the small area of first turn-on (less than 1 mm^2 in most structures) requires careful overload protection. A very severe condition, especially for this area, is the RC snubber-network discharge which results in nearly unlimited dI/dt rising to a current level of several tens of amperes. The turn-on stress of auxiliary thyristors can be partially removed by a series resistor, so that dI/dt is not as severe a demand as in optically triggered power thyristors.

Amplifying gate structures with relatively low amplification ratio (low ratio between the second stage to the first stage trigger current) enable very fast unloading of the first switched thyristor stage. The fast onset of amplification in the n$^+$p-n partial transistor is sufficient to produce an enhanced trigger current for the second stage. This point has been discussed by Voss,[25] Jaecklin[13] and also, in great detail, by Temple[15] who presented model calculations for structures with five amplification stages and amplification ratios between 2.5 and 5 (the first stage starting with 2 mA trigger-current sensitivity), which had very fast subsequent turn-on to avoid harmful temperatures in the device.

The turn-on behavior of thyristors can be well investigated by observation of the IR recombination radiation. Fig. 16 shows the turn-on of an amplifying gate-type auxiliary thyristor at V_D = 3.5 kV which was observed with a gated infrared image converter. We have also plotted the turn-on power. This device has a maximum blocking voltage of 4 kV, a minimum trigger power of 2.5 mW and its dV/dt capability at t_c = 80°C is above 4 kV/μs (V_{DM} = 2.6 kV). The circuit for the turn-on tests is shown in Fig. 17 (C_1 = 2 μF, C_2 = 3.4 nF, R_1 = 33 Ω, R_S = 33 Ω).

The recombination radiation pattern (1) was determined in the absence of load current, whilst pattern (2) was determined with a simultaneous load current. (1) shows the rest of the decaying trigger light in the 1.2 mm diameter trigger window. Already 1.5 μs after turn on (200 ns after turn-on delay) the second stage shows plasma luminescence, and at 2 μs, when the snubber discharge current is 40 A, the second stage has fully turned on. The turn-on ratio was considerably higher than in Temple's power thyristor (the calculated amplifying gate ratio may be in the range of 10 - 15), but in this auxiliary device, reliable turn-on behavior was obtained.

Although we have no direct experimental proof, it seems that in addition to the amplifying gate, field emitter action in the optically triggered thyristor is very helpful in protecting the first triggering area. Field emitter action is obtained when the cathode load current flows transversely in unmetallized n$^+$ emitter regions with relatively high sheet resistance.[26] This point is illustrated in Fig. 18 in which we show the lateral voltage difference obtained in a large area (6 mm diameter) n$^+$ emitter with re-

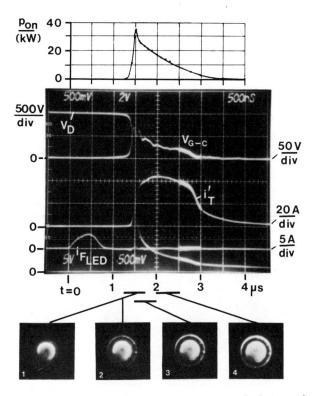

Fig. 16   Oscilloscope traces of turn-on power P_{ON} and time-resolved recombination radiation pattern illustrating turn-on of an amplifying gate-type auxiliary thyristor.

duced thickness (about 6 μm). The lateral voltage drop results in a transverse hole current in the p-base and this rapidly spreads the area which is first turned on. It seems that the most important effect is the turn-off action caused by the lateral current to this first area. This may be low in the first instants because of the high p-base sheet resistance, but probably increases once the injected carrier concentration modulates the conductivity. A similar voltage drop in the dV/dt compensated gate structure is caused by compensation resistor R_C.

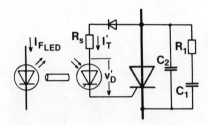

Fig. 17　　Test circuit for auxiliary thyristor turn-on tests.

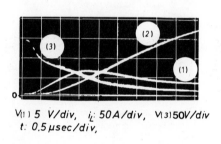

$V_{(1)}$ 5 V/div,　i_L: 50 A/div,　$V_{(3)}$ 50V/div
t: 0.5 μsec/div,

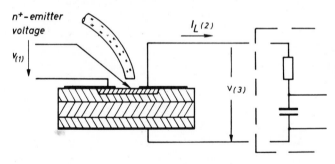

Fig. 18　n$^+$ emitter voltage enhancement due to field emitter action in a large optically sensitive gate (6 mm dia.).

An optically triggered amplifying gate structure which yields reasonably high dI/dt capability (250 A/μs at 2 kV, t_{vj} = 125°C) has been demonstrated by Ohashi et al.[27] for a 4-kV, 1.5-kA power thyristor. In this design, the pilot thyristor n^+ region is stepwise deep diffused and only partially covers the light-irradiated area in order to avoid turn-on at a single location.

Although it is more a problem of current capability than of dI/dt, a remark on the parallel operation of light-triggered thyristors should be made. All the devices discussed in this paper have integrated structures for high trigger-power amplification. This may impose limitations in parallel operation, since turn-on at very low anode voltage may not be possible (or only the pilot thyristor will turn on, but not the whole device). Turn-on of the whole device theoretically requires a voltage at least 0.6 V above the minimum on-state voltage of the pilot thyristor. In practice the value is much higher.

8. A SHORT SURVEY OF THE PRESENT STATUS OF DEVICE DEVELOPMENT

8.1 High-Voltage Auxiliary Thyristors

The development of high-voltage auxiliary thyristors has been discussed by Silber et al.[23] and Jaecklin and Bajan.[13] The advantages of auxiliary thyristors are:

a) One type of auxiliary thyristor can be combined with different types of normal electrical gate power thyristors.

b) Since auxiliary thyristors do not conduct a stationary load current, power dissipation and maximum operating temperatures are reduced. This improves the dV/dt capability significantly and enables the use of relatively simple device concepts. Because of fast unloading after turn-on, higher local switching power is tolerable.

c) Turn-off properties and forward-current capability are not important parameters. The design can be specialized to high optical sensitivity, blocking capability, dI/dt rating and dV/dt capability. Even a reverse-blocking capability is not always necessary.

d) Reduced demands for electrical and thermal contacts in the thyristor housing permit simple and efficient optical coupling systems.

Their disadvantages are:

a) Auxiliary thyristors are additional devices which must be designed for very high blocking voltages.

b) The possibilities for special gate concepts are restricted.

c) Triggering of the main thyristor requires a still higher minimum source voltage resulting in more limitations for parallel operation in auxiliary thyristor triggering than in direct optically triggered power-thyristor triggering.

Table 3 shows the most important published auxiliary thyristor data. The 3-kV device is a dV/dt-compensated device according to Fig. 13.

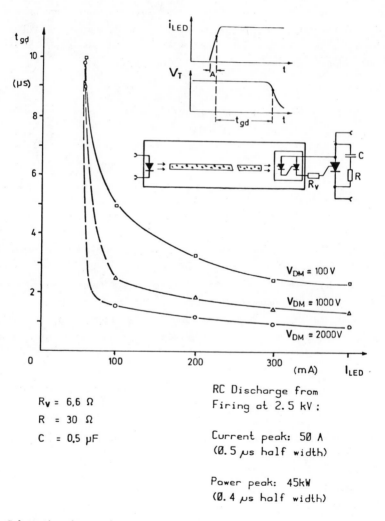

R_V = 6,6 Ω
R = 30 Ω
C = 0,5 µF

RC Discharge from
Firing at 2.5 kV :

Current peak: 50 A
(0.5 µs half width)

Power peak: 45kW
(0.4 µs half width)

Fig. 19   Schematic view and turn-on behavior of a high-voltage auxiliary thyristor photocoupler.

TABLE 3

Published Data on High-Voltage

Auxiliary Thyristors - LASCR's

V_{DRM}	I_{TAVM}	$(dV/dt)_{cr}$ at 70°C	P_{LT} at 25°C	References
4 kV	(∿ 4 A)	∿ 5 kV/µs	2.5 mW	1979 Jaecklin, Bajan[13]
4 kV	(∿ 5 A)	> 4 kV/µs	2.5 mW	
				1978 Silber, Füllmann, Winter[23]
3 kV	(∿ 5 A)	> 4 kV/µs	1.5 mW	

High-voltage auxiliary thyristors are also well suited to photocoupler-type compact devices. Fig. 19 shows the principle and the turn-on properties of a laboratory device with 10 cm glass-rod coupling. As discussed in section 2, the very efficient transmission properties yield good turn-on results with moderate LED current pulses.

TABLE 4

Published Data on Light-Activated

Power Thyristors - LASCR's

V_{DRM} kV	I_{TAVM} kA	min. trigger power mW	energy nJ	dV/dt at kV/µs	ϑ_i °C	References
4	1.5	2.5(device) 5 (at case)		1.5	125	1981 Ohashi et al.[27]
6	1.5	< 10		2	125	1980 Konishi et al.[11]
4	1.5	< 7		> 2	125	1980 Tada et al.[33]
2	1		35	> 1	125	1979 Page[31]
2.6	1		14	2	105	1979 Temple[32]
4.5	0.06	5		2	115	1977 De Bruyne et al.[16]
2.5	0.2	< 10		4	125	1977 Silber et al.[22]
2.5	0.2	1.5		> 1	125	

8.2 Light-Triggered Power Thyristors

Direct light-triggered power thyristors have been developed for different puroses and for triggering with different optical systems so that direct comparison of the data is not possible. The results, summarized in Table 4 show, however, that many development groups have obtained devices which satisfy high-voltage circuit application demands. Special methods for designing the optical coupling systems in the disk package are to be found in the patent literature. Interesting points include the technology of the glass-ceramic boundary[28] and special ways of obtaining elastic coupling and/or some additional light concentration by the use of glass cones.[29]

9. CONCLUSIONS AND SOME FURTHER ASPECTS

Research on the special problems of thyristor triggering with low gate current, and developments of light-activated high-voltage power and auxiliary thyristors have resulted in devices which can operate, even under severe conditions, using available optical components.

For common usage in high-voltage converter equipment, however, powerful devices with long-term stability must be available at a reasonable price. The GaAs-GaAlAs double heterostructure CW lasers and the heterostructure LED's seem to be the most promising devices.

Further development of such thyristors could be initiated by two aspects:

a) Further development of the integrated optical gate to obtain structures which exhibit safe breakover triggering (including dV/dt and reapplied dV/dt triggering) would be an obvious extension, since the problems of triggering at high voltage with low trigger power are similar to the problems of optical triggering. One example of a light-activated thyristor with safe breakover trigger capability has already been published.[3]

b) Technologies, such as the integration of MOS components in power devices, could perhaps be used for both fault-trigger protection and trigger-power amplification. Small-area, low-current planar triacs for line-voltage applications have already been developed.[20,30] MOS integration in high-voltage and high-power thyristors may, however, become a rather expensive technology. Nevertheless, a possible combination of these very different branches of silicon technology has considerable fascination for power-device designers.

REFERENCES

1. W. Gerlach and G. Köhl, Thyristoren für hohe Spannungen, Festkörperprobleme IX, Vieweg-Verlag, 1969, 354-370.

2. D. Silber and M. Füllmann, "Improved Gate Concept for Light-Activated Power Thyristors," Proc. IEDM, Washington D.C. (1975), 371-374.

3. P. De Bruyne and R. Sittig, "Light-Sensitive Structure for High-Voltage Thyristors," IEEE Power Electronics Specialists Conference (1976) Cleveland, Ohio, Conf. Report, 262.

4. V.A.K. Temple and P. Ferro, "High-Power Dual Amplifying Gate Light-Triggered Thyristors," IEEE Trans. Electron Devices, ED-23 (1976) 893-898.

5. D. Silber, W. Winter, and M. Füllmann, "Progress in Light-Activated Power Thyristors," IEEE Trans. Electron Devices, ED-23 (1976) 899-904.

6. E.S. Schlegel and D.J. Page, "A High Power Light-Activated Thyristor," Proc. IEDM, Washington D.C. (1976) 483-486.

7. J. Nakata, "Light-Activated Thyristor," U.S. Pat. 3 697 833 (Jap. 45/14600, 1970).

8. W. Gerlach, "Light-Activated Power Thyristors," 6th European Solid State Dev. Res. Conf. (ESSDERC), Solid State Devices (1976) 111-133.

9. D.E. Williamson, "Cone Channel Condensor Optics," J. of Opt. Soc. America, 42 (1952) 712.

10. D. Silber et al., "Lichtzündung von Leistungsthyristoren I," Arbeitsbericht BMFT 403-7291-NT 541 (1975).

11. N. Konishi et al., " A 6000 V, 1500 A Light-Activated Thyristor," Proc. IEDM, Washington D.C. (1980) 642.

12. H. Schlangenotto and W. Gerlach, "On the Effective Carrier Lifetime in p-s-n Rectifiers at High Injection Levels," Solid-State Electronics, 12 (1969) 267-275.

13. A.A. Jaecklin and I. Bajan, "Novel Gate Concept Improves Performance of Light-Fired Thyristor," Proc. IEDM, Washington D.C. (1979) 254-257.

14. V.A.K. Temple, "Directly Light-Fired Thyristors with High dI/dt Capability," Proc. IEDM, Washington D.C. (1977) 22-25.

15. V.A.K. Temple, "Comparison of Light-Triggered and Electrically Triggered Thyristor Turn-On," IEEE Trans. Electron Devices, ED-28 (1981) 860.

16. D. Kuse, P. De Bruyne, P.M. Van Iseghem, and R. Sittig, "New Voltage Limiters, Breakover Diodes and Light-Activated Devices for Improved Protection of Power Thyristors," 2nd Int. Conf. on Power Electronics - Power Semiconductors and Their Applications, IEE London (1977), 18.

17. D. Silber et al., "Lichtzündung von Leistungsthyristoren II," Arbeitsbericht BMFT 403-7291-NT 600 (1978).

18. P.M. Van Iseghem, "p-i-n Epitaxial Structures for High-Power Devices," IEEE Trans. Electron Devices, ED-23 (1976) 823.

19. P. Svedberg, Halbleiterordnung, Patent DE-AS P 26 25 917 (FRG), (750 7080, Sweden, 1975).

20. V.P. O'Neill, P.G. Alonas, and D.M. Gilbert, "A Monolithic Optically Isolated Zero Crossing Triac Driver," Proc. IEDM, Washington D.C. (1978) 107.

21. H. Patalong, Lichtzündbarer Thyristor, European Pat. No. 0 029 163 (DE 29 45 335, 1979).

22. D. Silber, M. Füllmann, and W.M. Lukanz, "Recent Developments in Light-Activated Power Thyristors," 2nd Int. Conf. on Power Electronics - Power Semiconductors and Their Applications, IEE London (1977) 14.

23. D. Silber, M. Füllmann, and W. Winter, "Light-Activated Auxiliary Thyristors," Proc. IEDM, Washington D.C. (1978) 575.

24. J. Burtscher, "Thyristoren mit innerer Zündverstärkung," VDE-Tagung, Dynamische Probleme der Thyristortechnik, Aachen (1971) 128-138.

25. P. Voss, Mit Licht steuerbarer Thyristor, Patent (FRG), DOS 25 38 549 (1977).

26. W. Gerlach, "Thyristor mit Querfeldemitter," Z. für angew. Physik, 19 (1965) 396-400.

27. H. Ohashi, H. Matsuda, T. Ogura, T. Tsukakoshi, and Y. Yamaguchi, "Directly Light-Triggered 4 kV - 1,500 A Thyristor (SL 1500 GX 21)," Toshiba Rev., No. 131 (1981) 19-22; and private communication.

28. M.H. Hanes and L.R. Lowry, Lichtgetriggertes Hochleistungshalbleiterbauelement (U.S. Pat. 800 706, 1977), DOS 28 08 531.

29. H. Ohashi and Y. Shirasaka, European Pat. 00 21 352 (Jap. 77 148/79).

30. J. Tihany, "Functional Integration of Power MOS and Bipolar Devices," Proc. IEDM, Washington D.C. (1980) 35.

31. J.K. Page, "Light-Triggered Thyristors for VAR Generator Applications," 7th IEEE/PES Conf. (1979) 222-226.

32. V.A.K. Temple, "Light-Triggered Thyristors for HVDC Applications," 7th IEEE/PES Conf. (1979) 213-221.

33. A. Tada, A. Kawakami, T. Miyazima, T. Nakagawa, K. Yamanaka, and K. Ohtaki, "4 kV, 1500 A Light Triggered Thyristor," Proc. 12th Conf. on Solid State Devices, Tokyo 1980, Jpn. J. Appl. Phys., 20 (1981) Suppl. 20-1, 99-104.

DISCUSSION
(Chairman: Dr. W. Gerlach, TU, Berlin)

W. Gerlach

Let me start with the following question concerning delay time in light-activated thyristors which would be shorter than the delay time of electrically triggered thyristors provided the overdrive in both cases is the same. Is there still a need to increase the light sensitivity of your thyristors to such high values, as both you and Dr. Sittig and Dr. Jaecklin did, in the special structures for light-activated thyristors?

D. Silber (AEG-Telefunken, Frankfurt)

I've shown two examples of the dependence of trigger-delay time against overdrive factor. What we found is a relatively small trigger-delay time at moderate overdrive factor and it seems that this is very different for devices made in different laboratories. Some of the slower devices seem to be typical and have less of a decrease in turn-on delay time for the same trigger sensitivity. In fact, we have a stronger decrease in trigger-delay time with overdrive factor than in normally triggered devices. Of course, in optically triggered thyristors you can have extremely fast turn-on if you trigger them with pulse lasers. Our own experiments with pulse lasers show that you can have a trigger delay in the range of 200 nanoseconds. However, as you can read in the paper by Temple,[1] you have to be careful to work with immediate firing of the parallel device by just creating enough carriers optically.

D. Crees (Marconi Electronic Devices Ltd., Lincoln)

You've described fairly complicated structures and so far as I've understood it they are mainly of auxiliary thyristor nature. How would you compare the future for auxiliary devices with the integrated approach that was described earlier? Do you see light-activated thyristors as always being an auxiliary structure or do you see the integration of your complicated structure into the large device?

D. Silber

I would say the auxiliary device should be the starting point for this technique for then you can use many different types of main thyristors triggered optically. The question of integrating the trigger sensitivity into the device is just a matter of how many devices you wish to bring to applications where they are triggered by light. If this is to become a general way of triggering high-voltage devices, then you should change to direct triggering of the large-area device. Of course, many problems of the gate design are easier in the auxiliary device where you can protect it by some additional circuitry. Also the temperature is lower.

H. Stemmler (BBC Brown, Boveri & Co. Ltd., Baden)

Do you know of any practical applications of such light-activated thyristors?

D. Silber

As far as I know, all laboratories which have worked in this field have so far only used them in test setups.

H. Irmler (Semikron GmbH, Nürnberg)

Have you tried to integrate the breakover diode and the light-activated thyristor?

D. Silber

Although this is an attractive property of the device it depends on the application as to whether it's really worth doing. Normally you need a very well-defined breakover limit which means that you have to use special breakover diodes. In general, I would say it's always a good property of the device to have this overvoltage trigger capability for if either dV/dt or overvoltage triggering occurs the device will survive.

A. Jaecklin (BBC Brown, Boveri & Co. Ltd., Baden)

You just mentioned that devices of this type are currently under test at various laboratories. What breakthrough would be required for them to go into practical application?

D. Silber

You have seen that the trigger limit of the device is typically in the range of a few milliwatts at the device. This can come directly from the hetero-structure laser, which is expensive, or from modern light-emitting diodes, which are both expensive and still under development. Once this trigger power is available at reasonable expense, and this will probably come, then the devices can be used just as they are now.

H. Melchior (ETH, Zurich)

How much trigger power will bring you to success?

D. Silber

About 5 milliwatts.

H. Melchior

That shouldn't be too hard.

D. Silber

No, but you need that for the long term and guaranteed, and 5 mW would give you a trigger overdrive factor of only 2, so you have to be careful that you can work with this trigger overdrive factor. However, it is possible: we have worked the devices in this range.

H. Melchior
From when on would you consider the cost of the light source to be acceptable?

D. Silber
That's a question more for the circuit engineers who set the total price and have to consider the competing solution of using indirect optical triggering which will be used so long as it's cheaper. 800 DM, for example, for a hetero-structure laser would be too much.

J. Tihanyi (Siemens AG, Munich)
You say that the triggering power is some milliwatts. What does that mean for practical application? How big is the current of the light-emitting diode?

D. Silber
That depends very much on the application. For example, for this photocoupler type as you've seen you have trigger currents of a few hundred milliamps. If you want this trigger power via large glass cables you can only use short pulses which must then be in the ampere region, and this works with normal good light-emitting diodes. Of course, one might then prefer using pulse lasers.

H. Stemmler
We have put into practice in one case such a light-activated auxiliary thyristor, but this is only in a protection unit which has to be fired when protection is needed. In all cases where periodic firing is required it is today not possible to provide the light power which is needed by such elements.

R. Jötten (TU, Darmstadt)
Could somebody comment on the news from CIGRE from the United States that prototypes are being built using light-activated auxiliary thyristors with compensators in the 10-20 MVA range?

P. Hower (Unitrode Corp., Watertown)
Westinghouse had a contract with EPRI using light-fired devices. It was not an auxiliary light source but the light cables came from LEDs which led directly to the devices and they used a glass rod to come down to the devices. The current and voltage should be reported in the EPRI contracts.

D. Silber
It's reported and published and the trigger level is about 35 nanojoule and for good turn-on it requires about 300 nanojoule, which is provided by laser.

REFERENCES

1.   V. Temple, "Inverter Light Triggering Thyristor with Unique Arm-Structure Amplifying Gate," IEEE Trans. Electron Devices, ED-28 (1981) 801.

M. Michel (TU, Berlin)

I wish to comment on the fact that reverse-conducting thyristors are also advantageous in applications using natural commutation. Today they are mostly used in converters with forced commutation.

Dr. Stemmler referred to two types of converter: the "voltage rectifier" and "voltage inverter". Referring to the source of the commutation voltage they can also be called "converters with a.c.-side commutation" and "converters with d.c.-side commutation", the principles of which are shown in Fig. 1a and Fig. 2a, respectively. These descriptions are very well known from the literature. Figs. 1b and 2b show examples of applications using load commutation and the Figs. 1c and 2c examples with forced commutation. Some of these circuits were also shown by Dr. Stemmler.

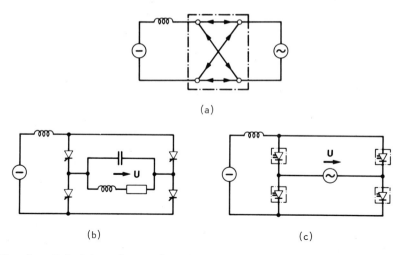

(a)

(b)   (c)

Fig. 1   Principle and examples of converters with a.c.-side commutation.

Simplifying still further we reach the principle circuits of the two types of converters presented in Fig. 3. They show that in both cases the converter circuit connects a current system with a voltage system. Fig. 3a is an example of a direct-current system (square symbol) connected by the converter to an alternating-voltage system (circle symbol), whilst Fig. 3b is an example of a direct voltage system connected by the converter to an alternating-current system. Due to the fact that for commutation a voltage is needed, the first example can be called "converter with alternating voltage

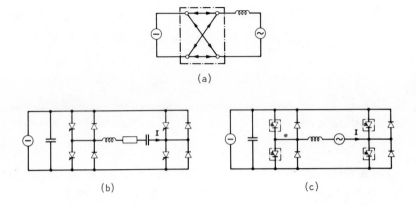

(a)

(b)   (c)

Fig. 2   Principle and examples of converters with d.c.-side commutation.

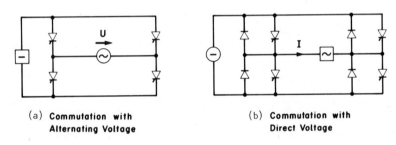

(a) **Commutation with**   (b) **Commutation with**
    Alternating Voltage   **Direct Voltage**

Fig. 3   Principle circuits of two types of converters.

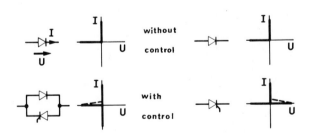

Fig. 4   Two types of valves.

commutation" and the latter "converter with direct voltage commutation". Other expressions might be "current converter" (Stromrichter) and "voltage converter" (Spannungsrichter).

Fig. 4 shows the two different types of valves used in the two types of converters. The converter with direct voltage commutation uses valves of the "voltage valve" type. This valve is characterized by two current directions and one voltage direction. The converter with alternating voltage commutation uses valves of the "current valve" type. The characteristic of this valve is marked by two voltage directions and one current direction.

Fig. 5 shows the circuit diagram of a six-pulse-bridge converter with natural direct-voltage commutation. A three-phase alternating-current system is connected to a di-rect-voltage system. Six valves of the voltage valve type are used. Some advantages of this type of converter in relation to the converter with natural alternating-voltage commutation can be seen, if consideration is given to the angle of overlap and to the control angle in the inverter mode. In Table 1 some formulae are given for the two types of converters. The following symbols are used:

$u_{a.c.}$ Angle of overlap (alternating-voltage commutation)
$u_{d.c.}$ Angle of overlap (direct-voltage commutation)
α Phase-control angle
β Control angle in inverter mode
γ Extinction angle
u_K Relativ commutation inductance

Figs. 6 and 7 show some results derived from the formulae of Table 1. Fig. 6 indicates the advantages of the converter with natural direct-voltage commutation in relation to the converter with natural alternating-voltage commutation.

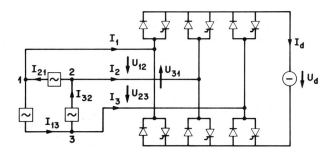

Fig. 5   Six-pulse-bridge converter with natural direct-voltage commutation.

TABLE 1

Commutation with

Alternating Voltage		Direct Voltage

Angle of Overlap

$u_{a.c.}$ $\qquad\qquad$ $u_{d.c.}$

$$\cos (u_{a.c.} + \alpha) = \cos \alpha - \frac{\pi}{3} u_K \qquad\qquad \sin (\alpha - u_{d.c.}) = -\sin \alpha + \frac{\pi}{2} \frac{u_{d.c.}}{u_K}$$

$$u_{a.c.} \underset{\sim}{} \frac{\pi}{3} u_K \frac{1}{\sin \alpha} \qquad\qquad u_{d.c.} \underset{\sim}{} \frac{4}{\pi} u_K \sin \alpha$$

$$u_K = \frac{\omega L_K}{\hat{U}_N / \hat{J}_N}$$

Extinction Angle γ
Control Angle (inverter mode) $\beta = 180° - \alpha$

$$\beta = \gamma + u_{a.c.} \qquad\qquad\qquad \beta = \gamma - u_{d.c.}$$

$$\cos \gamma = \cos \beta + \frac{\pi}{3} u_K \qquad\qquad\qquad \sin \gamma = - \sin \beta + \frac{\pi}{2} \frac{1}{u_K} (\gamma - \beta)$$

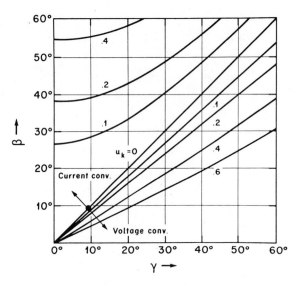

Fig. 6   Inverter control angle as a function of the extinction angle for the two types of converters.

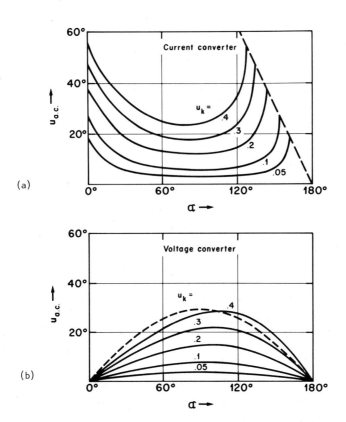

Fig. 7   Angle of overlap of the two types of converters
          a) Converter with alternating-voltage commutation
          b) Converter with direct-voltage commutation

W. Gerlach

We should now consider the question of future limits for the blocking voltage and current capability of power thyristors which represent the ratings of high-power thyristors. Has one now arrived at the technological limit of current capability with your large-diameter thyristors?

A. Jaecklin (BBC Brown, Boveri & Co. Ltd., Baden)

I would say that for the given diameter one has reached approximately a limit although there is no limit in an absolute sense. This means that if a larger diameter is available and if applications are there that need the larger current-carrying capability, there should certainly be the possibility of raising the limit. However, as I pointed out in the talk, according to the information we have from the applications side in the very near future the need seems to have a low priority.

H. Adler (General Electric, Schenectady)

Is there a practical upper limit to the voltage and how high a voltage device is anticipated in the near future?

A. Jaecklin

That's a good question which I should really have covered within the talk itself. If you review the literature you find that predictions of the upper limit have continually increased with time and technological advances, and this progress may well continue. The limits are essentially given by the fact that higher voltages require thicker crystals with correspondingly larger losses, longer turn-off time, larger recovery charges, and so on. It's just this optimization which eventually limits the practical voltage and it's my personal opinion that from the present point of view this limit has not been reached yet.

M. Adler

Aren't there thermal effects that begin to limit the further increase of blocking voltages? That's a real consideration with silicon. For high-voltage devices, the doping concentration falls below the intrinsic carrier concentration. Can you make some predictions on that basis?

A. Jaecklin

This again is somewhat related to the actual structure you're going to cope with and, as I have indicated, there is a possibility to go to asymmetric structures which automatically pushes you up by almost a factor of 2 with present type of structures. There may still be a further increase by as much as a factor of 2, but this may then be the real upper limit where you run into the problems you mentioned. However, with an asymmetric structure you have considerable freedom and a lot depends on the interaction with practical applications.

M. Adler

You mentioned free-floating silicon in your talk and the need for special handling techniques. Could you also comment on the relations to thermal resistance problems and longevity?

A. Jaecklin

The thermal-resistance effect was very nicely modelled in a paper in the IEEE Transactions on Electron Devices by yourself. We have made similar measurements because it is not an easy decision to change or go to a completely new concept. It's essentially a question of the time relationship because in the first few milliseconds you do have additional thermal resistance if you have just a plain pressure contact in contrast to an alloyed disc which does not have even microscopic air gaps in between. It's a matter of waiting long enough that the thermal resistances of both structures come to the same asymptotic value and this is estimated to be of the order of magnitude of the base width of the line-frequency cycle. Thus, it's just the transient for individual pulses which is somewhat impaired if you have a plain pressure contact. However, so far as the thermal properties over a longer range are concerned there is virtually no difference within the scope of interest.

M. Adler

You don't find that thermal fatigue is involved here?

A. Jaecklin

Our experiments so far indicate that the thermal fatigue with a plain pressure contact is a better solution than if you have a rigid contact which induces additional thermal stresses. The reason is that this rigidity has to be achieved at a fairly high temperature, the alloying temperature, which may be about 700° and which with normal thermal cycling may represent an additional stress to the silicon crystal.

H.-Ch. Skudelny (RWTH, Aachen)

In discussing power limits, not only the devices but also the circuitry should be considered. In this regard, what is the relative overcurrent capability of these devices and how can they be protected?

A. Jaecklin

For the specific devices we have discussed so far the overcurrent is approximately a factor of 7 - 8 above the rated current which makes it something more than 30 kA. There is, of course, the problem that with increasing voltage this ratio decreases. That means you do have a thicker tablet automatically so you have increased thermal effects with more heat dissipated. Since for surge currents you have essentially adiabatic conditions you're limited by the physics of heat generation.

GATE TURN-OFF THYRISTORS[*]

M. KURATA, M. AZUMA, H. OHASHI, K. TAKIGAMI,
A. NAKAGAWA, AND K. KISHI
Toshiba Research and Development Center, Kawasaki, Japan

SUMMARY

A number of aspects of gate turn-off thyristors (GTOs) will be discussed, including their historical background, technical demands requiring their development, operating principles, device modeling, design considerations, characterization, gate circuits, and a number of applications to actual equipment. The device design is implemented on the basis of modeling and test-sample experiments. The maximum gate turn-off current is shown to vary proportionally to the factors (V_{J1}/ρ_{SPB}) and W_{NB}, where ρ_{SPB} represents the sheet resistivity of the gated p-base layer, and W_{NB} the thickness of the nongated n-base layer. The use of a high doped p-base often leads to excessive reduction in the current-amplification factor of the n-p-n transistor portion, thus finally deteriorating the overall device performance. In extreme cases, it leads to abrupt increase in the on-state voltage, which is theoretically interpreted as resulting from bandgap narrowing. Representative characteristics are given for four types of GTO: 600 V - 200 A, 600 V - 600 A, 1300 V - 600 A, and 2500 V - 600 A units. Research and development activities at different organizations are also briefly reviewed. A number of GTO applications are illustrated, including an 18-kVA VVVF inverter for motor-speed control, a 170-kVA power supply for air conditioning, and a 610-kVA VVVF inverter for traction-motor control, both for electric railway coaches. A series of CVCF inverters from 30 kVA to 500 kVA output ratings is also mentioned.

1. INTRODUCTION

There have long been substantial requirements from the application field for a semiconductor device which is simultaneously provided with the blocking, triggering, and conducting characteristics of a standard thyristor together with the switching-off capability of a power transistor. Such a device would make it possible, in inverter and chopper applications, to eliminate forced-commutation circuits, thereby significantly contributing to reduction in equipment size and enhancement of power-conversion efficiency.

[*] Presented at the symposium by M. Kurata

The principle of the gate turn-off of a thyristor was discussed as early as 1960 by van Ligten and Navon,[1] who commented that the employment of an interdigitated emitter structure and reduction in either the n-p-n or p-n-p current gain are necessary conditions.

The theoretical analysis was extended by Wolley[2] in 1966 to investigate the gate turn-off transient, including discussion of the influence of several design parameters. He made a comparison between theoretical and experimental results for the storage time versus the turn-off gain characteristics, wherein test samples with gate turn-off current of more than 10 A were presented.

One of the earliest works on high power GTOs was published by New et al.[3] in 1970. They presented a developmental 1000 V - 50 A device and reviewed several factors, including the turn-off gain, the dynamic of junction recovery, and the gate voltage. Another report was given by Wolley[4] in 1973 in which a 1000 V - 200 A device was presented.

These early achievements stimulated successive movement toward GTOs with extended voltage-current ratings. Our own group started research and development activities in this period, first by trying to develop a numerical CAD model for GTOs based on charge-control equations, with the intention of simulating an entire set of anode-current waveforms in the time domain.[5] This model was used to investigate the influence on turn-off characteristics of fluctuations in the thickness of the p-base layer, and of carrier lifetimes in the p-base and the n-base layers. This model will be discussed in section 2.

At the same time research efforts were continued which ultimately yielded three types of high-power GTOs rated at 600 V - 200 A, 600 V - 600 A, and 1300 V - 600 A in 1977.[6,7] Later, the largest unit among our products, the 2500 V - 600 A type, was added in 1978,[8] together with advances in device-design concepts. From a device design point of view, methods of realizing a sufficient gate turn-off current I_{ATO} were investigated during the course of the work mentioned, and efforts made to retain those properties which are inherent in the standard type of thyristor. It was known that employment of a high-conductivity gated p-base layer contributes significantly to the increase in I_{ATO}. However, excessive increase in the p-base conductivity frequently led to an abrupt increase in the on-state voltage at a medium current level, thus imposing a serious limitation on the high-power GTO design. These topics will be covered in section 3. From a device application point of view, it is absolutely essential to have a basic knowledge of the overall device performance, as is discussed in section 4, including frequency characteristics.

It was mentioned previously that requirements from the application side promoted the development of high-power GTOs. At the same time, a kind of overexpectation seems to have existed among the device users, at least in the early stages. Frequently, the GTO was expected to be a device with blocking, triggering, and conducting characteristics comparable with standard type thyristors, with the additional gate turn-off operation to be achieved by as weak a signal as that required for triggering. Obviously, this was not the case in practice. Rather, it was found through theory and practice that those characteristics of the GTO which are inherent in the standard thyristor are generally inferior, or at best equal, to those realizable with the latter.

Seemingly, many application engineers were so accustomed to standard type thyristors that they were unwilling to consider sacrificing any of their excellent properties, such as high triggering and dV/dt capability. However, thanks to developments in device-design methods, GTO properties have gradually been accepted among application engineers in such a reasonable manner that appropriate design of the gate circuit and the snubber circuit was made. Numerous examples of GTO equipment are illustrated in section 5, both developmental and also commercial, thus verifying GTO capability as a real new product.

2. GTO MODELING

For the purpose of modeling the GTO, we consider the basic device configuration illustrated in Fig. 1, thereby assuming that the GTO is initially conducting. When the switch S is closed, a reverse-gate bias is applied to prevent electrons from being injected in such an area that the lateral voltage drop along the gated p-base layer is less than the applied voltage $E_{GC} - R_G I_G$. The rest of the entire device, if available, continues to conduct current even with the reverse-gate bias. Therefore, in order to achieve gate turn-off for the entire device, the lateral resistance along the p-base layer should be so low that the reverse bias becomes effective in the portion most remote from the gate electrode.

An overall simulation for the gate turn-off transient can be implemented by introducing charge-control equations and a hybrid two-dimensional modeling concept. The basic building block is a thyristor segment to be equivalently represented by a two-transistor analogue as illustrated in Fig. 2. Each transistor is described by a charge-control equation including currents to and from the neighboring segments:[5]

$$\frac{dQ_1(N)}{dt} + \frac{Q_1(N)}{\tau_{B1}} = I_{C2}(N) + I_1(N+1) - I_1(N), \tag{1}$$

$$\frac{dQ_2(N)}{dt} + \frac{Q_2(N)}{\tau_{B2}} = I_{C1}(N) + I_2(N) - I_2(N+1). \tag{2}$$

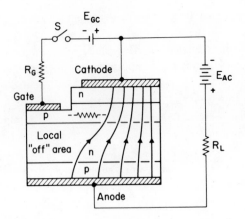

Fig. 1   Basic GTO configuration.

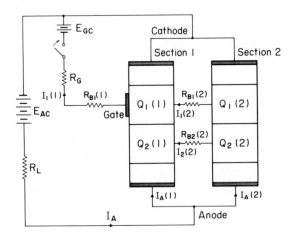

Fig. 2   Two-section GTO model.

The section number is denoted by N, excess charges in the p-base and n-base by Q_1 and Q_2, and base time constants by τ_{B1} and τ_{B2}, respectively.

In general, three different conditions per section are involved in the gate turn-off process, i.e., a saturation condition, a nonsaturation condition, and a depletion condition. Equations (1) and (2) take appropriate forms depending on these conditions.

In addition, an equation to describe the relation between charge and voltage is required for each of the three junctions. One such equation is given for the center junction J_2 as,

$$V_{J2}(N) = \sqrt{[\frac{f(N)}{2C_{J2}(N)}]^2 + V_{J20}^2} - \frac{f(N)}{2C_{J2}(N)} , \tag{3}$$

where

$$f(N) = Q_2(N) + \frac{\tau_{C2}}{\tau_{C1}}[\frac{Q_1(N)}{2} + \sqrt{[\frac{Q_1(N)}{2}]^2 + Q_1^0(N)^2}] - g(N), \tag{4}$$

$$g(N) = \frac{\tau_{C2}I_{J2}(N)}{2} + \sqrt{[\frac{\tau_{C2}I_{J2}(N)}{2}]^2 + Q_1^0(N)^2}. \tag{5}$$

Junction capacitance and current are denoted by C_{J2} and I_{J2}, respectively; V_{J20} and Q_1^0 are fitting parameters; and τ_{C1} and τ_{C2} are collector time constants which are related to τ_{B1} and τ_{B2} by,

$$\tau_{C1}/\tau_{B1} = (1-\alpha_{npn})/\alpha_{npn}, \quad \tau_{C2}/\tau_{B2} = (1-\alpha_{pnp})/\alpha_{pnp}. \tag{6}$$

Finally, a number of current loop equations are defined to complete the GTO model. Interconnection resistances R_{B1} and R_{B2} are assumed to include the conductivity modulation by a relevant excess charge as given by,

$$R_{B1}(N) = R_{B10}(N) \frac{Q_{10}(N)}{Q_1(N)+Q_{10}(N)}, \tag{7}$$

$$R_{B2}(N) = R_{B20}(N) \frac{Q_{20}(N)}{Q_2(N)+Q_{20}(N)} , \tag{8}$$

where quantities with suffix zero represent those corresponding to the low injection level.

The current state of the art has permitted substantial improvements to one-dimensional thyristor modeling.[9] In addition, hybrid two-dimensional thyristor modeling has reached a higher level of development so that more rigorous expressions can be given for junction I-V characteristics, current-transfer characteristics, and the capacitances representing excess charge storage.[10,11] A more advanced simulation is now possible employing these improved concepts.

3. DEVICE DESIGN CONSIDERATIONS

One of the main problems of high-power GTO design is how to equip a four-layer thyristor structure with the gate turn-off capability without losing its inherently excellent properties, such as high blocking voltage, high gate triggering sensitivity, low conduction voltage, and high surge-current capability.

In the field of heavy industrial electronics, there has been a desire that a GTO incorporate all the possible properties of a thyristor in addition to the gate turn-off capability. In fact, very high power, low-frequency equipment has traditionally been built with large high-power thyristors. For such applications, a low on-state voltage is especially important, because power dissipation in the steady conduction state represents most of the total loss. Therefore, a thyristor-like GTO is needed.

On the other hand, medium power, high-frequency equipment has been built with either power transistors or small high-speed thyristors with low switching losses, because overall power dissipation is strongly influenced by switching losses during the turn-on and turn-off phases. Therefore, fast switching and high turn-off gain are important. In other words, a transistor-like GTO is desirable. In this paper, emphasis is given to the former type, according to requirements of heavy industry applications.

3.1 Factors Determining the Maximum Gate Turn-off Current I_{ATO}

Investigation of the maximum gate turn-off current, I_{ATO}, is a basic aspect of the design of GTOs of any kind. During the last ten years, I_{ATO} values have been reported in the 50 to 600 A range, and 1000 A has recently been reported.

In a one-dimensional sense, a thyristor can generally be switched off by applying a reverse-gate bias, if both of two transistors are driven out of saturation. In a two-dimensional sense, the voltage drop along the semiconductor surface should not exceed the applied voltage across the gate and the cathode electrodes.

Qualitatively, the former condition also holds for a standard type thyristor, because otherwise it cannot be turned off by either the reverse-gate bias or the forced commutation. On the contrary, the latter condition is only peculiar to the GTO, because the lateral drop required to turn off a standard type thyristor will be much in excess of the allowable gate voltage. These considerations lead to the conclusion that a GTO must be composed of many narrow emitter elements.

The most dominant factor in determining I_{ATO} is undoubtedly the localized current crowding which occurs in a few elements during the gate turn-off process. In the following, the current crowding is investigated for single cathode samples or samples of small size, including model calculation results and basic experiments.[12]

3.1.1 Model Calculation Results

The CAD model for a GTO as described in section 2 is employed to investigate the influence of several important design parameters. Figures 3(a) and (b) show anode-current waveforms obtained from a two-section model for two different p-base concentrations, (a) $C_{PB} = 3 \times 10^{17}$ cm^{-3}, and (b) $C_{PB} = 3 \times 10^{18}$ cm^{-3}. Other input parameters are specified in Table 1.

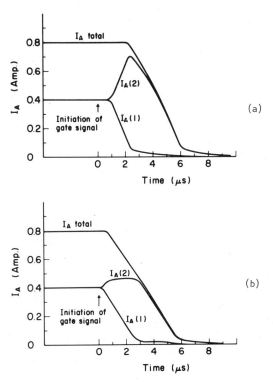

Fig. 3   Computed anode currents during gate turn-off (a) $C_{PB} = 3 \times 10^{17}$ cm^{-3}, (b) $C_{PB} = 3 \times 10^{18}$ cm^{-3}.

In the former case, the current crowding is much more marked than in the latter, because the high resistivity p-base hinders removal of excess carriers out of the gate electrode. In other words, p-base conductance is one of the most important factors in determining the crowding, and thus also I_{ATO}. This is in agreement with what was quantitatively discussed in section 2.

TABLE 1

Model Input Parameters

(See Fig. 12 for notation)

Doping parameters:

$C_{NE} = 4\times10^{20}$ cm^{-3}, $C_{NB} = 2\times10^{14}$ cm^{-3}, $C_{PE} = 3\times10^{18}$ cm^{-3}

$W_{NE} = 10$ μm, $W_{PB} = 50$ μm, $W_{NB} = 99$ μm, $W_{PE} = 60$ μm

Lifetimes:

$\tau_{B1} = 5$ μs, $\tau_{B2} = 1.5$ μs

Lateral Sizes:

Cathode width = 100 μm, gate width = 120 μm,

Spacing between cathode and gate = 30 μm

Cathode and gate finger length = 4 mm

External circuit conditions:

$E_{AC} = 600$ V, $E_{GC} = 10$ V

$R_{L} = 750$ Ω, $R_{G} = 62.5$ Ω

3.1.2 Basic Experiment 1: Emitter Pattern Variation

Even for conditions of fixed impurity distribution, the efficiency of p-base conductance can be increased by employing a fine cathode patterning. Therefore, variation in emitter finger width will finally influence I_{ATO}. Such experiments were made on a group of test samples with results given in Fig. 4, where line (a) corresponds to a single element, and (b) to several elements with a constant total emitter area of 0.7 cm^2. In both cases, the element shape is rectangular with a constant length of 7 mm. These results show an increase in I_{ATO} with decrease in the finger width. It should be noted that the multielement I_{ATO} values are only twice as large as those for a single element. For a width of 0.8 mm, I_{ATO} is around 40 A for one element, while it is only 80 A for 12 elements. This suggests a localized nonuniform anode-current distribution in a few elements.

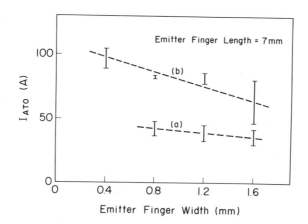

Fig. 4   Experimental I_{ATO} vs. emitter finger width characteristics (a) single element, (b) multiple elements.

The nonuniformity is thought to occur even in a single element switching. Figures 5(a) and (b) show anode-current waveforms for a single element and two elements, respectively. In the latter, the total current I_A begins to localize in element C_1 accompanying its increase beyond a certain limit. Consequently, a nonsmooth change in I_A

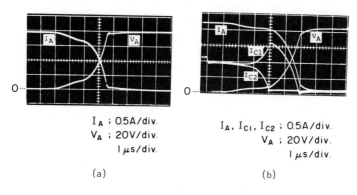

I_A ; 0.5A/div.
V_A ; 20V/div.
I μs/div.

(a)

I_A, I_{C1}, I_{C2} ; 0.5A/div.
V_A ; 20V/div.
I μs/div.

(b)

Fig. 5   Gate turn-off waveforms (a) single element, (b) two elements.

appears during the storage phase. Depending on the situation, a similar shape can be observed even in the single-element case (a), suggesting that the conduction area in the element breaks up into several islands.

3.1.3 Basic Experiment 2: Infrared Microscopic Technology

Electrical measurements discussed above are supplemented by infrared microscopic technology,[12] which enables direct observation of the current nonuniformity in a single element.

It is known that two modes of radiation recombination exist in high carrier-injection conditions, one of which is attributed to recombination through traps located midway in the bandgap, to yield photon radiation of around 2.2 μm wavelength. The other process is due to direct recombination of electron-hole pairs, resulting in 1.1 μm wavelength radiation. Since the latter occurs proportionally to the excess electron-hole pair density in the space-charge neutral zone, it yields information about the inner carrier density.[13] Fig. 6 illustrates the system for infrared observation which has a partial resolution restricted by the precision of the optical system to 20 μm.

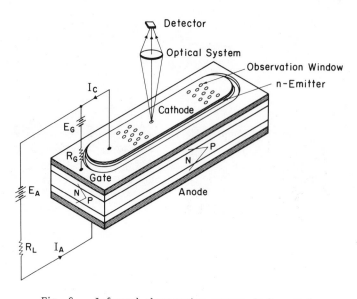

Fig. 6   Infrared observation system (schematic).

Many windows of 30-µm diameter are formed on the cathode metal to detect the infrared emission. Fig. 7 shows the anode current and voltage with carrier-density distributions at four sampled times t_1 through t_4.

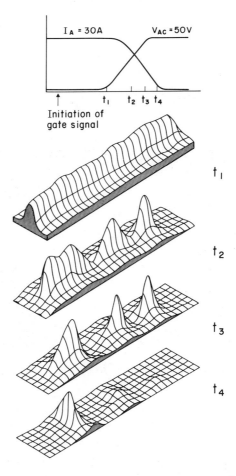

Fig. 7   Current and voltage waveforms and carrier distributions.

From the initiation point of the gate signal to a late period in the storage phase t_1, conduction region squeezes only in the width direction without any essential change along the length direction.

From t_1, the conduction region squeezes along the length direction, to break up into several peaks. The squeezing proceeds further, finally to form a single peak at t_4.

Such nonuniformity in carrier density is thought to result from either the carrier lifetimes or p-base sheet resistivity. The above results suggest that to achieve high gate
turn-off capability, an excessively long emitter finger is undesirable.

3.1.4 Experiment on Full-Size Devices

As a result of the aforementioned basic investigations, a cathode emitter pattern was
determined for a 600 A I_{ATO} unit as given in Fig. 8, whereby consideration was also
given to various limitations on material processing technologies. It consists of 260 fingers, each having 0.3-mm width and 4-mm length. The total emitter area is about
3 cm^2.

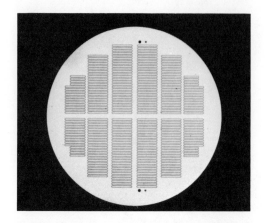

Fig. 8   Mask pattern for 600-A GTO unit.

Fig. 9 shows an experimental I_{ATO} versus (V_{J1}/ρ_{SPB}) characteristic, where the gate
cathode junction breakdown voltage is denoted by V_{J1}, and the p-base sheet resistivity by ρ_{SPB}. Thickness and resistivity for the n-base are kept constant. The approximately linear dependence is in accordance with the expression,

$$I_{ATO} = K \cdot G_{off} \cdot V_{J1}/\rho_{SPB}, \tag{9}$$

where K is constant for a fixed emitter pattern, and G_{off} is an operational turn-off
gain. Generally, G_{off} is considerably less than the maximum gain, $G_{off}(max)$, and is
mainly determined by the external gate circuit.

In addition, there are experimental data for the dependence of I_{ATO} on the n-base
thickness W_{NB}, which show that I_{ATO} increases linearly with W_{NB}.[8] This is probably
due to a quasi balance-resistance effect exhibited by the n-base layer. Therefore, a
high-voltage GTO unit with a thick n-base is advantageous in achieving a high I_{ATO}.

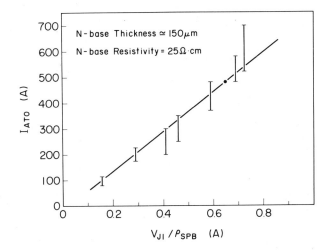

Fig. 9   Experimental I_{ATO} vs. (V_{J1}/ρ_{SPB}) characteristic.

3.2 Forward-Blocking/Conduction Characteristics

In addition to the gate turn-off capability, an overall design tradeoff is required among the other characteristics which are inherent in a standard type thyristor. These include the forward-blocking voltage, the on-state voltage, the surge-current capability, the latching and holding currents, the turn-on and turn-off times.

Among these quantities, we investigate the forward-blocking voltage and the on-state voltage. A key factor in the determination of these characteristics is the carrier lifetime in the p-base layer. In particular, a high lifetime in the p-base contributes to improvement in the voltage-blocking capability through reduction in the center junction leakage current. In addition, it contributes to improving the overall GTO characteristics through enhancement in α_{npn}.

Increase in the p-base lifetime is achieved by introducing some appropriate annealing process in the course of the fabrication process. One such example is given in Fig. 10, where a second phosphorus deposition is introduced before the gold diffusion process to reduce the n-base lifetime. Improvement is demonstrated for 2500-V units in Fig. 11, where the leakage current versus the off-state voltage characteristics are given for pre- and post-introduction of the annealing. At the same time, the on-state voltage was improved from 3.3 V to 2.2 V, the surge current capability from 3000 A to 5000 A, and the latching current from 10 A to 2 A.

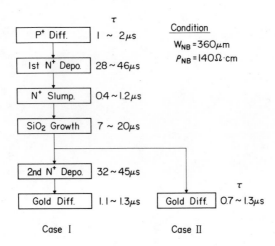

Fig. 10   Process flow chart and carrier lifetimes at each stage.

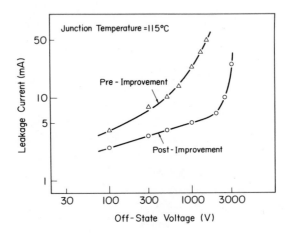

Fig. 11   Leakage current vs. off-state voltage characteristics at 115°C junction temperature.

Another important factor which influences the overall thyristor characteristic is an acceptor concentration at the cathode gate junction prior to n-emitter diffusion. This parameter is denoted by C_{PJ1}, as illustrated in Fig. 12. During the course of research on increasing I_{ATO}, it was found that an excessively high acceptor concentration of the p-base layer led to an abnormal forward I-V characteristic as shown in Fig. 13.

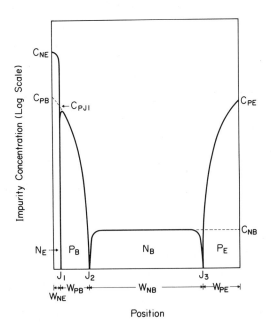

Fig. 12   Impurity doping profile for GTO.

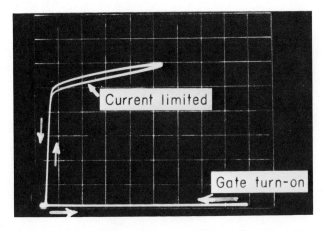

Fig. 13   Experimental anode current vs. voltage characteristic.

The CRT trace shows an abrupt increase in the on-state voltage at relatively low anode current, as though the GTO behaves like a transistor in its nonsaturation, active operating condition.

Measurements by a probing method made of the potential distributions along the bevelled sidewall of a real device show that most of the voltage drop exists across the center junction.[14] The samples listed in Table 2 were prepared and tested to yield the critical current versus C_{PJ1} characteristics given in Fig. 14.

TABLE 2

Doping Conditions and Critical Anode Current I_{crit} Characteristics
for Fabricated GTO's

(W, C, and I are μm, cm^{-3}, and A, respectively)

	W_{NE}	W_{PB}	W_{NB}	C_{NE}	C_{PB}	C_{NB}	C_{PJ1}	I_{crit}
A	9.2	43.3	150	5×10^{20}	2.8×10^{18}	2.8×10^{14}	2.1×10^{18}	22- 40
B	11.6	41.9	150	4×10^{20}	2.8×10^{18}	2×10^{14}	1.8×10^{18}	100-300
C	13.3	40.7	150	3×10^{20}	2.8×10^{18}	2×10^{14}	1.5×10^{18}	> 1800
D	15.4	39.8	150	2.5×10^{20}	2.8×10^{18}	2×10^{14}	1.2×10^{18}	> 1800

In addition, α_{npn} was shown experimentally to decay linearly with increase in C_{PJ1} beyond about 10^{18} cm-3. It follows that the singularity is due to reduction in α_{npn} caused by excessively high C_{PJ1}.

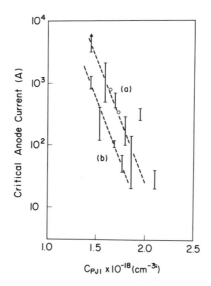

Fig. 14   Critical anode current vs. C_{PJI} characteristics with gold diffusion temperature as parameter. (a) below 840°C, (b) 860 to 880°C.

An effort has been made to investigate the phenomena described by means of a theoretical bandgap narrowing model.[15] Computed I-V characteristics given in Fig. 15 show
that bandgap narrowing, rather than Auger recombination or carrier-to-carrier scattering, is the dominating factor for the singularity. It was shown that the model
leads to semiquantitative order of agreement between theory and experiment. In addition, it was deduced that the singularity can be avoided by keeping α_{npn} in excess
of 0.73.

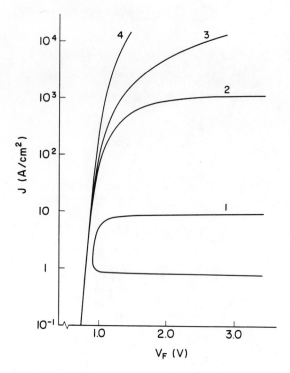

Fig. 15   Theoretical anode current vs. voltage characteristics: 1. Heavy doping
           effects (HDE), Auger recombination (AR), and carrier-to-carrier scattering
           (CCS) included; 2. HDE excluded; 3. HDE and AR excluded; 4. HDE, AR,
           and CCS excluded.

4. DEVICE CHARACTERIZATION

Representative characteristics for the 1300 V - 600 A unit are listed in Table 3, with
test conditions for each item. Evaluation of I_{ATO} was made by the resistive load,
chopper circuit given in Fig. 16.[6] Measurements showed that there is an appreciable
dependence of I_{ATO} on external circuit conditions, i.e., the snubber circuit and the
gate-drive conditions, which are specified in the figure.

TABLE 3

Main Characteristics for a 1300 V - 600 A GTO

Test Item	Typical Value	Test Conditions
Maximum gate turn-off current	600 A	See Fig. 16: $T_j = 110°C$
Turn-on time	11 μs	$I_G = 3$ A, $V_{AK} = 650$ V
Delay time	5 μs	$t_G = 20$ μs, $T_j = 25°C$
On-state voltage	1.8 V	$I_A = 600$ A, $T_j = 25°C$
Gate turn-off time	12 μs	See Fig. 16: $T_j = 110°C$
Min. d.c. gate trig. current	300 mA	$V_{AK} = 24$ V, $T_j = 25°C$
Min. d.c. gate trig. voltage	0.63 V	$V_{AK} = 24$ V, $T_j = 25°C$
Surge on-state current	6000 A	50 Hz one-cycle peak
Gate-cathode breakdown voltage	14 V	$T_j = 25°C$
Holding/latching current	3 A	$T_j = 25°C$, $I_G = 3$ A, $t_G = 20$ μs, $R_L = 5$ Ω
Off-state voltage	1300 V	$T_j = 115°C$, $R_{GK} = 20$ Ω

Note: t_G = gate pulsewidth

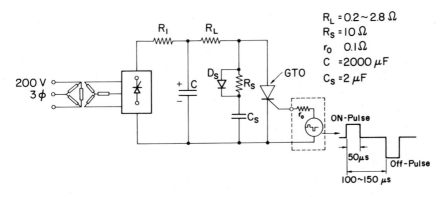

Fig. 16   Chopper circuit for evaluating gate turn-off characteristics.

Typical switching waveforms are given in Fig. 17 for the gate turn-on and turn-off transients. The turn-on time typically rated at 11 μs, is apparently limited by α_{npn}.

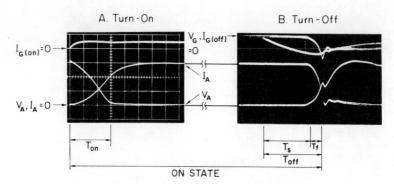

Fig. 17   Switching waveforms, V_A = anode voltage (200 V/div); I_A = anode current (200 A/div); I_G = gate current (10 A/div for turn-on, and 120 A/div for turn-off); V_G = gate voltage (20 V/div); time scale = 2 μs/div.

Numerous test results show that the anode current rise rate increases with increase in α_{npn}. The turn-off time, rated at 12 μs, is separated into the storage time and the fall time, with the former occupying the greatest fraction of the total time. The storage time depends on the on-state current and the rate-of-rise of gate current (dI_G/dt), which is 30 A/μs in the given example.

Operational turn-off gain, defined as $(I_A/I_{G,peak})$ also depends on (dI_G/dt). The gain is around four at (dI_G/dt) = 30 A/μs, and between 6 and 8 at (dI_G/dt) = 10 A/μs. However, trying to enhance the gain by reducing (dI_G/dt) leads to an increase in the turn-off time, and thus also the gate power dissipation.

An additional point of interest is the choice of the gate source voltage amplitude E_G. According to a basic experiment on a two-segment GTO unit, it was found that a high E_G (typically in excess of the static breakdown voltage V_{J1}) provides a desirable condition towards well-balanced simultaneous turn-off. A typical recommendation for a 2500 V - 600 A unit is E_G = 30 V, which is twice V_{J1}. It is a matter of course that the high voltage must be terminated at the appropriate time, to avoid excessive gate power dissipation.

With respect to the snubber circuit, a spike in the anode voltage appears in the fall time phase, due to a stray inductance. In order to minimize the spike, it is necessary to locate the snubber-circuit components closely to the GTO unit.

From a device application point of view, the limitation of GTO operating frequency is an important problem. This is determined by the total power loss involved in turn-on and turn-off, and the steady on-state, wherein the former two are determined by the product of frequency and the respective switching-power dissipation, while the latter is determined by the product of duty cycle, anode current, and on-state voltage.

Taking the 2500 V - 600 A unit as an example, these factors were determined on a semiexperimental basis to yield a number of temperature versus frequency characteristics as given in Fig. 18, under the assumption of a resistive load chopper circuit with a 50 percent duty cycle and square wave operation. For 1 kHz operation, I_{ATO} is predicted to be 400 A for a 70°C case temperature.

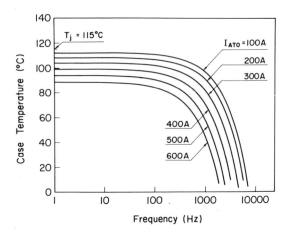

Fig. 18   Theoretical case temperature vs. frequency characteristics with I_{ATO} as parameter.

In summary, the main characteristics are listed in Table 4 for the four types: 600 V - 200 A, 600 V - 600 A, 1300 V - 600 A, and 2500 V - 600 A. In particular it should be mentioned that those items inherent in standard type thyristors, such as low on-state voltage, low latching current, and high surge-current capability are all kept at sufficient levels.

Apart form the above results, a number of recent activitites concerned with high-power and high-speed GTO development have been carried out in various organizations. With respect to high-power devices, one design principle is based on employment of the anode-short structure, whereby diffusion of gold as a lifetime killer is excluded.[16,17] Another design principle is based on the employment of a low-high impurity

TABLE 4

Main GTO Characteristics

Item	600 V 200 A	600 V 600 A	1300 V 600 A	2500 V 600 A
Off-state voltage	600 V	600 V	1300 V	2500 V
Max. gate turn-off current	200 A	600 A	600 A	600 A
On-state current	140 A	400 A	400 A	300 A
Turn-on time	7 μs	11 μs	11 μs	11 μs
Delay time	4 μs	4 μs	4 μs	4 μs
On-state voltage	1.6 V	1.8 V	1.8 V	2.2 V
Gate turn-off time	10 μs	13 μs	13 μs	18 μs
Operational gain	3	3	3.5	4
Surge on-state current	2500 A	6000 A	6000 A	5000 A
Holding/latching current	2 A	2 A	2 A	2 A

concentration structure for the n-base to improve an overall design tradeoff, whereby the reverse-blocking capability is considered nonessential.[18] Both of these approaches are reported to have successfully yielded 2500 V - 1000 A units.

With respect to high-speed GTOs, a 1200 V - 160 A unit is reported to have achieved an operating frequency of 50 kHz by employing diffusion of iron prior to that of gold.[19] Another report describes an extensive study on high-speed GTOs including several ideas for improving the device performance.[20] These are, (1) optimized control of gold distribution, (2) a series inductance in the gate circuit, (3) Schottky barriers instead of the usual anode shorts, (4) resistivity ballasted cathodes with the ohmic contact existing only at the periphery.

To summarize all the above examples, GTO characteristics cover a wide range of voltage, current, and speed. Including a variety of structural modifications, appropriate choice of design conditions makes it possible to equip a GTO with the high power-handling capability of a large standard type thyristor, or the high-speed performance of a transistor in the same category. In addition, design conditions so determined are mostly realizable without great difficulty through the standard device-fabrication technology at the present state of the art, even considering the production yield problem which is indispensable to commercial products.

5. APPLICATIONS

In spite of a history of more than a decade, GTOs are still in a sense new devices, especially concerning application to high-power equipment, which are built using standard type thyristors. However, GTOs are no longer exploratory or developmental devices, but commercial products to be employed with the substantial advantage over standard type thyristors that massive forced commutation circuit components are not needed. This will be verified in the following through a number of actual examples. For further details and further applications, reference may be made to a recent publication dealing specifically with GTOs and GTO equipment.[21]

5.1 18-kVA VVVF

The first example is one of the earliest products among GTO equipment designed for the speed control of induction motors.[22] The variable voltage, variable frequency (VVVF) inverter is characterized by the following data:

Installed GTOs	600 V - 200 A (6)
Input	3 ϕ, 200 V, 50 Hz
Output	3 ϕ, 18 kVA at 200 V, 1000 Hz
	Voltage range: 20-200 V
	Frequency range: 100-1000 Hz
	Instantaneous overload: 170 %

5.1.1 Main Circuit

Fig. 19 shows the main circuit configuration consisting of a standard type thyristor bridge unit for the output-voltage control, and a GTO inverter unit for the output-frequency control. The former portion is composed of six 600 V - 130 A thyristors, each of which is controlled by an independent gate circuit. The latter is composed of six 600 V - 200 A GTOs, each of which is accompanied by a diode to feed back the inductive energy to the d.c. power source. In addition, each GTO is provided with an independent gate circuit and a snubber circuit. The latter is to reduce the steep dV/dt which appears in the gate turn-off and is often referred to as the "commutation dV/dt".

5.1.2 Gate Circuit

The gate circuit plays an essential role for operating the GTO under appropriate conditions. Fig. 20 shows the gate circuit employed for the present VVVF. Each of the

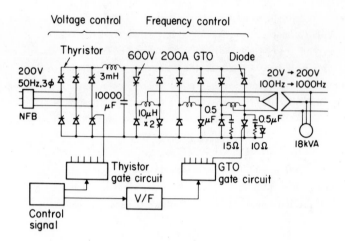

Fig. 19   18-kVA GTO-VVVF, main circuit.

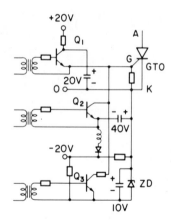

Fig. 20   18-kVA GTO-VVVF, gate circuit.

six gate circuits is fed by an independent d.c. power source of $\pm$ 20 V. There are three transistors of which Q_1 is for the turn-on, Q_2 for the higher gate current at the initial part of the turn-off, and Q_3 for the lower gate current to be fed during the tail part in the anode current. The latter is intended to achieve a high dV/dt capability at a quasi or perfect off-state, sometimes referred to as the "static dV/dt".

The turn-on current pulse has a 6-A peak followed by a flat zone of 0.5 A. The turn-off current pulse has a 90-A peak with dI_G/dt = 20 A/μs, followed by a lower current phase with more than 10 μs duration. For the rest of the off-state, a negative voltage is applied to prevent the GTO from misfiring. Recently, an improved version of the gate circuit was developed which substantially reduces the total power dissipation.[23]

5.1.3 Operation

The speed up and reduction operation on induction motors is controlled by a V/F-converter in Fig. 19, which has a linear frequency versus voltage characteristic, in the range of 20 V to 200 V, and 100 Hz to 1000 Hz. The rise and fall times from 100 Hz to 1000 Hz, and vice versa, can be varied in the range of 10 to 990 minutes by adjusting digital switches in the front panel.

5.2 Static Power Supply for the Air Conditioning of Railway Coaches

Fig. 21 shows the circuit diagram of a 170-kVA power supply which is composed of a thyristor chopper and two sets of three-phase GTO inverters to form a twelve-phase bridge configuration.[24] The 1500-V d.c. line voltage is supplied through a pantograph. The output a.c. power is fed to battery chargers, fluorescent lamps, fans, and air-conditioning units.

The thyristor chopper, of the series-commutated type, involves four 2500 V - 400 A reverse-conducting thyristors. It generates a pulse-modulated voltage across the intermediate d.c. link LC tank circuit. The chopper elementary frequency is kept as high as 360 Hz, so that the output d.c. voltage may be assumed to be flat.

Each three-phase GTO inverter consists of a group of three half-bridge inverters which yield rectangular voltage waveforms with control shifted by 120°. Two sets of three-phase inverters are shifted in phase by 30° from each other and coupled together by inverter transformers to provide a twelve-phase voltage waveform. This configuration makes it possible to obtain a quasi-sinusoidal output voltage.

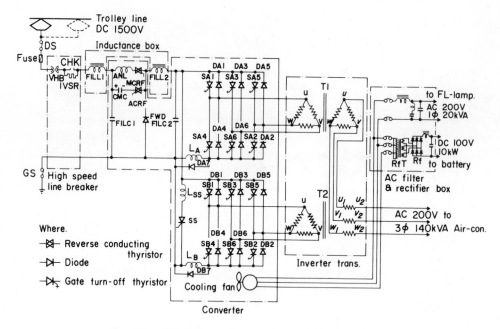

Fig. 21   170-kVA static power supply, main circuit.

Advantages of the GTO power supply over one utilizing standard type thyristors are summarized as follows:

1)   Reduction of noise by 3 to 5 dB (absolute value 69 to 73 dB).
2)   Equipment weight reduced by 400 kg. Reduction in area occupancy reduced by 10 %.
3)   Distortion in the a.c. output waveform reduced from 11.5 % to 6.7 %.
4)   Efficiency enhanced by 2 % to achieve 84 % in total.

All these improvements are consequences of the removal of the forced-commutation circuits which became possible by introducing GTOs instead of standard type thyristors.

There is another more recent product of the same category which is rated at 50 kVA, 3 phase, 200 V. The total circuit configuration is further simplified compared to the above type, thus resulting in further improvement in the noise level of 65 to 70 dB, reduction in weight by 12 % relative to a rotating-machine system, and power-conversion efficiency of 92 %.

5.3 VVVF for Traction Motor-Control of Railway Coaches

This equipment is one of the most recent products among electric-railway applications. Traction-motor control is built using twelve GTO's of 2500 V - 800 A type. This GTO unit is a powered up version of the 600-A type.

The output capacity is 610 kVA, with the load of two induction motors of 160 kW each, and the d.c. line voltage is 750 V. The frequency-control range is from 5 to 125 Hz. The equipment is characterized not only by its high-power capacity, but also by the introduction of a microcomputer system for pulsewidth modulation and frequency control.

5.4 TOSNIC-1000 Series CVCF

Apart from the railways applications, there are substantial demands for constant-voltage, constant-frequency converters to provide, in connection with batteries, noninterruptable, regulated a.c. power supplies for high-reliability systems, such as high-performance computers, broadcasting stations, airport control terminals, and medical equipment in hospitals.

In order to meet these demands, a new TOSNIC (= TOShiba NonInterruptable Converter) - 1000 series was developed by installing 1300 V - 600 A or 800 A GTO units. This series ranges in output from 30 kVA to 500 kVA. Taking as an example the TOSNIC-1400 series of 200 kVA, the total efficiency is 94 %, which exceeds a previous standard type by 4 %. Improvement is also achieved in floor-area occupancy and weight, by 48 % and 32 %, respectively. In addition, noise is reduced by 10 dB to 65 dB. All these results are due to removal of the forced-commutation components.

6. CONCLUSIONS

Discussion has been given of high-power GTOs with respect to their modeling, device-design considerations and characterization, wherein not only their performance, but also limiting factors were described. To summarize, present design, fabrication, and application technologies have advanced so far that GTOs are not only interesting for research and development, but have proved themselves to be commercial products to be employed with substantial advantages over standard type thyristors in a variety of applications, as described through actual examples.

REFERENCES

1. R.H. van Ligten and D. Navon, "Base Turn-Off of p-n-p-n Switches," IRE WES-CON Convention Record, Part 3 on Electron Devices (August 1960) 49-52.

2. E.D. Wolley, "Gate Turn-Off in p-n-p-n Devices," IEEE Trans. Electron Devices, ED-13 (July 1966) 590-597.

3. T.C. New, W.D. Frobenius, T.J. Desmond, and D.R. Hamilton, "High Power Gate-Controlled Switch," IEEE Trans. Electron Devices, ED-17 (September 1970) 706-710.

4. E.D. Wolley, R. Yu, R. Steigerwald, and F.M. Matterson, "Characteristics of a 200-Amp Gate Turn-Off Thyristor," IEEE Conference Record, Industrial Applications Society Meeting (1973) 251-255.

5. M. Kurata, "A New CAD-Model of a Gate Turn-Off Thyristor," IEEE Power Electronics Specialists Conference Record (1974) 125-133.

6. M. Azuma, A. Nakagawa, and K. Takigami, "High Power Gate Turn-Off Thyristors," Proceedings of the 9th Conference on Solid State Devices, Tokyo, 1977, Japanese Journal of Applied Physics, 17 (1978) Supplement 17-1, 275-281.

7. H. Ohashi, M. Azuma, and T. Utagawa, "High Voltage, High Current Gate Turn-Off Thyristor," Toshiba Review, 112 (November/December 1977) 23-27.

8. M. Azuma, K. Takigami, and M. Kurata, "2500 V, 600 A Gate Turn-Off Thyristor (GTO)," IEEE International Electron Devices Meeting, 246-249; IEEE Trans. Electron Devices, ED-28 (March 1981) 270-274.

9. M. Kurata, "One-Dimensional Calculation of Thyristor Forward Voltages and Holding Currents," Solid-State Electronics, 19 (1976) 527-535.

10. M. Kurata, "Hybrid Two-Dimensional Device Modeling," Proceedings of the NASECODE-II Conference, Boole Press, Dublin, 1981.

11. M. Kurata, Numerical Analysis for Semiconductor Devices, Lexington Books, D.C. Heath and Company, Lexington, Mass., 1981.

12. M. Azuma, M. Kurata, and H. Ohashi, "Design Considerations for High Power GTO's," Proceedings of the 12th Conference on Solid State Devices, Tokyo, 1980, Japanese Journal of Applied Physics, 20 (1981) Supplement 20-1, 93-98.

13. R.A. Kokosa, "The Potential and Carrier Distributions of a p-n-p-n Device in the ON State," Proceedings of the IEEE, 55 (August 1967) 1389-1400.

14. M. Azuma and K. Takigami, "Anode Current Limiting Effect of High Power GTO's," IEEE Electron Device Letters, EDL-1, (October 1980) 203-205.

15. A. Nakagawa, "Numerical Analysis on Abnormal Thyristor Forward Voltage Increase Due to Heavy Doping in Gated p-Base Layer," Solid-State Electronics, 24 (1981) 455-459.

16. T. Nagano et al., "High-Power, Low-Forward-Drop Gate Turn-Off Thyristor," IEEE Conference Record, Industrial Applications Society Meeting (1978) 1003-1006.

17. S. Sakurada, H. Matsuzaki, and Y. Ikeda, "Recent Trends of Gate Turn-Off Thyristors," (in Japanese), Hitachi Hyoron, 63 (June 1981) 369-372.

18. A. Tada et al., "GTO with 2500 V Off-State Voltage and 1200 A Controllable Anode Current," (in Japanese), General Meeting of Japanese Institute of Electrical Engineers (1981) Paper No. 446.

19. H. Hayashi, T. Mamine, and T. Matsushita, "A High-Power Gate-Controlled Switch (GCS) Using New Lifetime Control Method," IEEE Trans. Electron Devices, ED-28 (March 1981) 246-251.

20. H.W. Becke and R.P. Misra, "Investigations of Gate Turn-Off Structures," IEEE International Electron Devices Meeting (1980) 649-653.

21. Toshiba Review, Special Issue on GTO's and GTO equipment (in Japanese), August 1981.

22. K. Kishi, M. Kurata, K. Imai, and N. Seki, "High Power Gate Turn-Off Thyristors and GTO-VVVF Inverter," IEEE Power Electronics Specialists Conference Record (1977) 268-274.

23. N. Seki, Y. Tsuruta, and K. Ichikawa, "Gating Circuit Developed for High Power GTO-Thyristors," IEEE Power Electronics Specialists Conference Record (1981) 215-225.

24. T. Jinzenji, F. Moriya, T. Kanzaki, and M. Azuma, "Three-Phase Static Power Supplies for Air-Conditioned Electric Coaches Using High Power GTO," IEEE Trans. Industry Applications, IA-17 (March/April 1981) 179-189.

ACKNOWLEDGEMENT

Grateful acknowledgement is made for permission to reprint the following figures: 2, 3, 4, 5, 6, 7, 9;[12] 8, 10, 11, 18;[8] 12, 13, 14;[14] 15;[15] 16, 17;[6] 19, 20;[22] and 21.[24] Table 2 is also reprinted here by permission.[14]

DISCUSSION
(Chairman: Prof. H.-Ch. Skudelny, RWTH, Aachen)

P. Zimmermann (Robert Bosch GmbH, Erbach)
You showed the 600 A-type GTO. Do you run it with a constant gate current like a transistor? Also, you showed the snubber network with a resistor, a diode and a capacitor. Normally this is not possible because when you switch on the lower GTO you always charge the upper capacitor directly. Can you just discharge the capacitors with the GTO and what are the restrictions in that field?

M. Kurata (Toshiba R&D Center, Kawasaki)
The example I used for illustration is just for a testing unit. Our application engineers have a snubber circuit incorporating just capacitance and resistance, but, I think, no diode.

P. Zimmermann
Can you just discharge the capacitor with the GTO without using a resistor which causes an additional voltage drop?

M. Kurata
Without a resistor, the voltage spike will appear very soon after turn-off and the device will be destroyed, so I don't think that's good.

W. Gerlach (TU, Berlin)
What do you think about the implementation of anodic shorts to improve turn-off behavior?

M. Kurata
Many manufacturers are interested in using an anode short structure. I have nothing against introducing it, but one can do without it. The anode short structure is not the determining factor, but comes in the third or fourth place. The other factors are more important in achieving reasonable overall characteristics.

A. Jaecklin (BBC Brown, Boveri & Co. Ltd., Baden)
You have not mentioned anything on the future prospects for GTOs. Does this mean that the units you mentioned like 2500 V, 1000 A somehow represent a limiting design or do you expect further increase and improvement of these devices in the future?

M. Kurata
I don't think that 2500 V, 1000 A is a limit. We can go up to higher ratings depending on the requirements of the application.

D. Schröder (Universität Kaiserslautern, Kaiserslautern)
You told us that the GTO is sensitive to overcurrents especially when you switch off. Do you have some more protective devices or something else?

M. Kurata

In some cases our application engineers introduce a protecting circuit composed of a usual thyristor whilst in other cases they have no usual thyristors but a self-consistent GTO system with a protection function. A consequence is that the user must have good understanding of the device.

CONVERTERS WITH HIGH INTERNAL FREQUENCY

F.C. SCHWARZ
Delft Unversitiy of Technology, Delft, The Netherlands

1. INTRODUCTION

Continuing work on the development of the technology of electric-power converters indicates the desirability of relatively high internal operating frequencies in order to satisfy a number of advantageous system characteristics. It appears desirable to equip
these power converters with a number of properties which are summarized in the following paragraph.

A power converter can be viewed as a device which derives electric energy from an
available source and transforms it into the replica of an input signal. Ideally no trace
of the form and shape in which the original electric energy is available should remain,
and this energy should assume the form and shape of the replica of an input or control signal. This control signal could be available at a very low power level, e.g. milliwatts, and replicate with a scaled magnitude and with the associated dynamics.

The reproduction of a control signal at a higher level of energy corresponds to the
function of an ordinary amplifier. Here, however, this "amplifier" does not enjoy the
luxury of being fed with well-controlled power. Also desirable is a high level of efficiency. Another desirable characteristic is that the physical size and weight of the
converter should be kept as low as possible, especially when it is meant to be port-
able or mounted on a vehicle. The limitation of physical size and weight also has a
significant impact on the cost of the equipment, because of the reduced use of materials. A further aim is directed toward the improvement of functional properties.
These can include two- or four-quadrant operation of the system as viewed at any of
its terminal pairs. The road toward all these goals is facilitated concurrently with the
application of principles and techniques which are well known to us in association with
communication technology.

The analogy between a power converter and an amplifier suggests the use of the terminology used in communication technology. This analogy is directed toward the area
of communication theory concerned with the use of pulse-modulation techniques for the
transfer of signals. The transformation of electric energy into a train of pulses is
used for the shaping and control of this energy on its way to a load area. The repe-

tition rate of these pulses is limited by the characteristics of the switches involved in the pulse-forming networks. These characteristics include the heat dissipation in the switches, which is co-determined by the associated network characteristics.

The pulse-control processes referred to above do not enjoy the customary benefit of being solely subject to more simple techniques, such as pulsewidth modulation, unless relatively heavy filters are used to restrict the rates of change of voltages and currents. If it is desired to achieve relatively fast response, and thus avoid heavy filters, then the form and the shape of each individual pulse cannot be immediately predicted. More sophisticated methods must then be applied for the pulse-forming processes.

2. METHODS OF PULSE FORMATION; POWER DISSIPATION AND FREQUENCY

Two of the significant pulse-forming methods, as used in power converters, are briefly reviewed here in order to explain the significant aspects of power-pulse-modulation processes. Fig. 1 indicates a basic presentation of the "chopper" circuit, which is used to implement pulsewidth modulation processes.

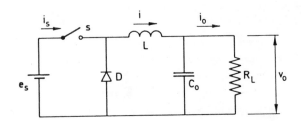

Fig. 1   Symbolic schematic of a power pulsewidth modulator.

A source current I_s is derived from a source with voltage e_s. This current I_s is switched on and off by repeatedly opening and closing switch S. The current I flowing through the inductor L, is indicated in Fig. 2(a).

During the time interval of commutation of the current I from being conducted by switch S and the diode D, we find transition periods in which both elements form a short circuit across the source of energy or find themselves in various states of transition varying between full conduction to "open" circuit, and vice versa.

The dissipation P_{dw} in a semiconductor switch, represented by S, is indicated schematically in Fig. 2(b). Most of the dissipation occurs during the time of transition when

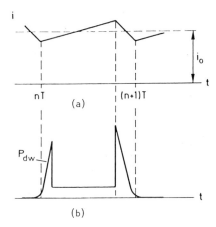

Fig. 2   Typical waveforms of power pulsewidth modulators: (a) the inductor current
I; (b) the power dissipation P_{dw} in the switching element S.

the switch S is turned on or off. The extent of this dissipation, and its effect on the
total switch dissipation $P_{d\ av}$, is the main cause of frequency limitations of chopper
circuits. This limitation is usually found to be between 200 and 400 Hz. A consequence
is that the number of pulses per second could not be increased without an unacceptable increase in the heat dissipation in the semiconductor switching elements S and D.
This limitation is largely due to the switching phenomena rather than the heat dissipation caused by conduction losses, which are little affected by the switching transients.[1]

Another method has been considered which achieves a process of pulse modulation
without imposing as severe a penalty on the pulse-forming process as that described
above. The principles of this process of pulse modulation are discussed with reference
to Figs. 3 and 4. Equivalent circuits of the power system which characterize the recurrently succeeding states, are depited in Fig. 3(a) and (b). To the left is a d.c.
source of energy and to the right the filter-capacitor d.c.-motor parallel combination,
analogous to that shown in Fig. 1. However, the switching mechanism which controls
the transfer of current from the source to the load though is now different.

The switch itself, consisting of a thyristor Th_i and antiparallel diode D_i (i = 1,2), is
placed in series with an LC combination, as shown. Conduction of the current I_1 by
the thyristor Th_i is necessarily terminated whenever the series capacitor C_1 achieves
the potential that will terminate the current. The resonant current I_1 then flows towards the source e_s through diode D_i until such time that further changes are imposed. At this point the requirements of the equivalent circuit change from the configuration Fig. 3(a) to 3(b), or vice versa. Again, the thyristor takes the lead and

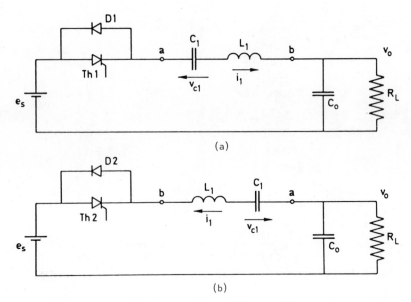

Fig. 3   Equivalent circuits of the converter with resonant circuits: (a) $I_1 > 0$; (b) $I_1 < 0$.

carries the current I_1 until its resonant turn-off, as described above. The returning current is, again, carried by the antiparallel diode until such time that the previous equivalent configuration takes over.

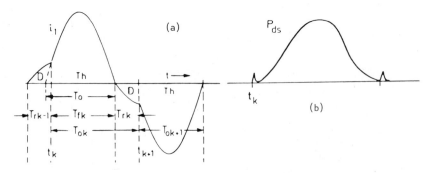

Fig. 4   Typical waveforms of resonant current power-pulse modulators: (a) the quasi-resonant current I_1; (b) the power dissipation P_{ds} in the switching thyristor.

The resulting current I_1 is indicated in Fig. 4(a). The thyristor and diode currents are identified as portions of the resonant current I_1 in the time T_{fk} and T_{rk}, respectively. Detailed treatment of this method of power-pulse generation is found in the literature.[2]

A simplification was used in the presentation of the equivalent circuit shown above. It is contended here that the resonant LC circuit can be momentarily disconnected, reversed and reconnected into this equivalent configuration so that the terminals a and b are interchanged every time the configuration a and b are interchanged. Various actual circuits can be used to realize this principle.

The average of the absolute value of I_1 per unit of time is recorded by an integrator which then controls the firing mechanisms of the thyristors. The excitation frequency of the resonant circuit is below its own natural frequency. The obvious effect is that further lowering of this excitation frequency leads to an increase in the apparent impedance between source and load and results therefore in a reduced transfer of power to the load. The resulting control of power transfer, together with its quantitative implications, is described in the literature.[2]

The important feature that results from the process of power transfer described above is the mechanism and the associated switch stress patterns. Fig. 4(b) indicates the power dissipation in a thyristor when operating in the equivalent circuits shown in Fig. 3. Turn-on occurs in the presence of the intended inductance, which limits the rate of rise of current. During its initiation phase the current is limited by the inductance L_1. Conduction can thus be established in a predictable and current-limited condition. Termination of the current occurs when all inductive elements in series with the switch are devoid of energy storage, because the current I_1 is on the point of reversal. The power dissipation at the initiation and at the termination of the current pulse carried by the thyristor is therefore very moderate, especially when compared to the power dissipation at the fringes of the current pulse indicated in Fig. 2(b).

Power dissipation in thyristors, as described above, consists essentially only of the conduction losses inherent in semiconductor devices. The power loss is therefore almost "independent" of the thyristor's frequency of operation. The frequency of operation of systems comprising of circuits which are equivalent to those of Fig. 3, is now solely limited by the physical capability of the semiconductor devices to follow their control signals. This is in contrast to the type of circuits depicted in Fig. 1, where the power dissipation which is created by the succession of switching events, is the limiting factor. The option of employing systems such as those shown in Figs. 3 and 4, creates problems which are based in the control area and will be discussed in the following section.

3. PULSE MODULATION IN RESONANT-CURRENT SYSTEMS

The system described above with reference to Figs. 3 and 4, employs the controlled and time-varying excitation frequency of a resonant circuit for the purpose of control of the flow of energy. The control system is therefore necessarily that of an aperiodic pulse mechanism. Moreover, there is no definable carrier frequency, around which specific variations of frequency occur. The current-wave oscillations in the resonant LC circuit consist of a sequence of current pulses which are not defined a priori. The individual current pulses are defined as lasting from the moment of ignition of one thyristor at time t_k to the ignition of the next thyristor at time t_{k+1}. This time interval

$$t_{k+1} - t_k = T_{ok} \qquad\qquad\qquad (1)$$

varies with the variations of source voltage and the degree of loading. The load terminal receives charge of one polarity, which is equal to the absolute value of I_1.

The control system is so devised as to form each current pulse in accordance with the average value of a current-control signal during the same time interval. The current-control signal I_r is related to I_1 by the following integral equation

$$\int_{t_k}^{t_{k+1}} I_r \, dt \sim \int_{t_k}^{t_{k+1}} I_1 \, dt; \qquad I_r(t) \neq I_1(t) \qquad\qquad (2)$$

The moments of thyristor ignition are expressed by

$$t_k = \sum_{m=0}^{k-1} T_{om}; \qquad\qquad T_{om} \neq T_{on} \qquad\qquad (3)$$

The sequence of current pulses which reaches the load terminal filter combination, consists therefore of current pulses that are not necessarily equal in amplitude or duration and their individual average value follows that of the reference signal I_r as determined by equation (2). If I_r were constant, then the output of the system would be a constant current. The system can be converted to a controlled output-voltage system by application of the well-established techniques of control engineering. It can be said in summary that the relatively simple concept of pulsewidth modulation control has been transformed here to a method of pulse-area control, which is necessarily more complex. It is the price paid for regaining the freedom of termination of substantial current pulses with a moderate effort of dissipation in the switching device.

The extent to which it is possible to generate low-frequency power waveforms depends on the pulse frequency to harmonic content ratio, which will govern the power-pulse modulation process involved. In order to generate a 50-Hz sine wave with an adequate degree of "high fidelity" would require a pulse frequency to power frequency ratio in excess of 10, and if possible a ratio of 100. A pulse frequency f_F of at least

$$f_F > 100 \times 50 = 5 \text{ kHz} \tag{4}$$

would be required in this case.

A turn-off time of approximately 10 μs has been associated so far with thyristors with a current-carrying capability up to some kiloamps. The current waveform I_1 shown in Fig. 3(a) could be interpreted as indicating that the diode current phases contain the intervals during which the thyristors are being turned off. Experience and considerations of the controlability of the relevant pulse train have shown that the ratio of duration of thyristor pulses versus the diode current under the conditions shown, could comfortably lie within the limits of 3 to 5. This means that one current pulse with duration T_{ok} could have a duration of 50 μs, corresponding to a pulse frequency of 20 kHz. Converters of that kind with power capacities in the order of 100 kW and measured efficiencies in excess of 96 % have been built and tested.[3,4]

4. WAVE SHAPING

The pulse-modulation process described above, and the associated bandwidth aspects, indicates that it is possible to convert a.c. voltage or current waveforms to d.c. or other forms of a.c. voltage. This includes loads such as d.c. or a.c. machines and the transformation of electrical power at certain interface points of utility grids with asynchronous tendencies.

The electrical energy enters a three-phase network terminal of the converter, is converted to a high-frequency "carrier" and then decomposed to follow two, three or more terminals of one or several ports.

In essence we have the situation whereby the principles of classical communication theory are applied in a fundamental form, and whereby the pulses are governed by principles of pulse area versus time ratio rather than being controlled in their width, as is otherwise usual.

The result of the phenomena described, associated with a pulse-forming process, is that the pulse intervals follow with unequal duration. A control system, which is governed by equation (2), still guarantees that the train of power pulses which is being generated, contains the character of the control signal and that the output of the converter will consist of a replica of the control signal I_r. The characteristic waveforms of the converter's output, which at times resemble a sequence of unipolar sinusoids, are smooth when high-frequency filters are being applied with cut-off frequencies below the lowest applicable pulse-frequency rate. The nominal maximum pulse frequency could amount to 20 kHz; the lowest practical pulse-frequency range is then still in the region of so many kilohertz.

Voltage scaling is achieved within the high-frequency operation of the converter, which allows the application of transformers for operating frequencies of the order of 10 kHz. Transformers of this kind require iron core materials which are more expensive than customary transformer iron. Also, the specific core losses are substantially higher, even with the use of high quality magnetic materials. It is fortunate that the higher frequencies also reduce the necessary cross-sectional area of the transformer core, so that the transformers as a whole are ultimately less expensive than conventional transformers. In addition they are substantially smaller and lighter than their 50-Hz counterparts.

5. THE DC CONVERTER

The principles of the d.c. to d.c. converter of this type are reviewed first in order to present the more simplistic form of this class of converters. A simplified schematic of a converter, which transfers energy from a source of d.c. power with voltage e_s to a load in form of the armature of a d.c. machine, is illustrated in Fig. 5.[5]

The two ports of the converter, connected to the source of energy and to the load, are shown in the form of bridges of semiconductor switches. These two bridges would be completely equivalent if the diodes D_{ij} (i,j = 1,2) were to be replaced by

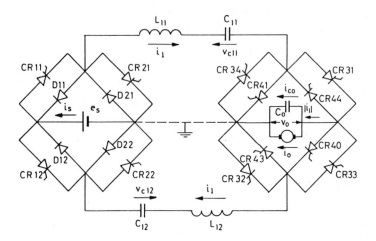

Fig. 5   Series resonant circuits for a four-quadrant d.c. drive.

thyristors. The filter capacitor C_o, which parallels the load, constitutes a short circuit for the high-frequency content of the rectified resonant currents. These currents travel through the two halves of one resonant circuit which consists of a series inductor $2L_{1i}$, and a series capacitor $\frac{1}{2}C_{1i}$ (i = 1,2). The thyristor pairs CR_{11}, CR_{22} and CR_{12}, CR_{21}, are energized in succession to close the circuits which include the series resonant parts described earlier. Simultaneous conduction of the thyristor pairs CR_{31}, CR_{32} and CR_{33}, CR_{34}, will represent a condition whereby the source with voltage e_s is connected to the load, as described before. The antiparallel diodes D_{ij} (i,j = 1,2) and the corresponding thyristors CR_{4i} in the load bridge, will perform the functions which bring about a current as indicated in Fig. 4(a).

A general system presentation and the associated current waveforms are summarized in Fig. 6.

It is seen here that the two converter "halves" comprising the switching matrices SM_1 and SM_2 respectively, would be entirely symmetrical if the diodes D_{ki} (k = 1,2; i = 1,2,3,4) were replaced by thyristors. A schematic of the various forms of the power-carrying current is depicted in Fig. 6(b).

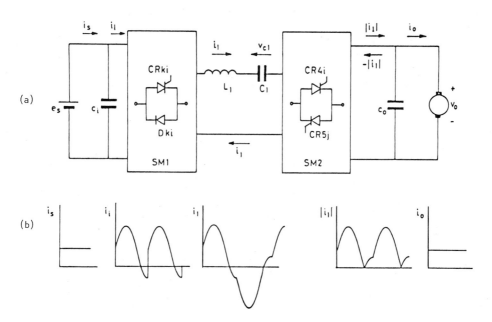

Fig. 6 Functional schematic of the four-quadrant drive: block diagram, the significant current waveforms.

A more detailed schematic of the various converter-current waveforms is shown in Fig. 7. Also indicated in this figure is the net "driving voltage" V_{LC}, which is impressed on the resonant circuit and causes the generation of the current waveform I_1. The significance of the current I_i leaving the input capacitor C_i and the current in the resonant circuit is revealed when comparing parts (a) and (c) of Fig. 7. Fig. 7(c) also indicates the manner in which energy is derived from the source of electric energy and suggests the principle of power control by maintaining the proper relation between the positive and negative parts of the current I_i. For instance, ideally,

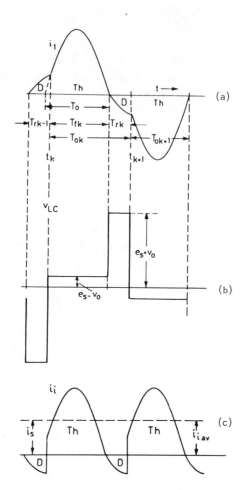

Fig. 7   Current waveform (a) I_1 in the resonant circuit; (b) the "driving voltage" V_{LC} of the resonant circuit; (c) the current I_i entering the switch matrix SM_1 near the source terminals.

$$\int_{t_k - T_{rk-1}}^{t_k} I_i dt = -\int_{t_k}^{t_{k+1} - T_{rk}} I_i dt \tag{5}$$

if the output terminals of the converter are short circuited and "zero power" is drawn from the source of electric energy. In this case the converter rejects the acceptance of power from the source of electric energy.[2]

A simple single-quadrant d.c. converter is shown in Fig. 8. The power circuit of a converter as discussed above and comprising a resonant circuit, is shown in the upper part of this figure. A block diagram of the associated control and protection system is attached to the power circuit mentioned.[6]

For the sake of convenience of presentation only one half of the bridge circuit is shown.

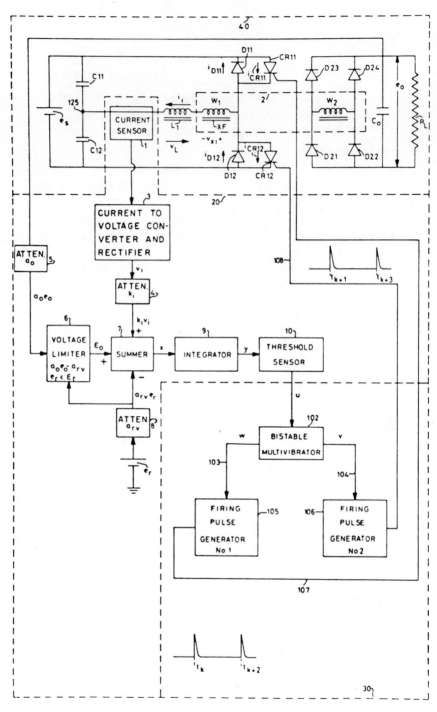

Fig. 8   Single-quadrant d.c. converter system comprising resonant circuit.

6. THE SECOND QUADRANT OF OPERATION

Return of electric energy from the "output bridge" to the "input bridge" of Fig. 5 is, in general, referred to as second-quadrant operation.

The "load current" I_0 then assumes a negative polarity, as shown in Fig. 9, so that

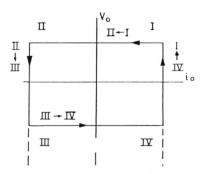

Fig. 9   The four quadrants of operation as viewed from the output terminals.

the positive output voltage V_0 and the negative $-I_0$ form a negative product. A load, such as a d.c. motor, then "produces" energy, which is fed back into the source of electric energy with voltage e_s. The motor voltage decays in the process of "discharge" of the motor energy. A decaying motor voltage V_0 has to keep the resonant circuit working while operating in a condition where

$$|V_0| < e_s \tag{6}$$

The "net driving voltage", discussed with reference to Fig. 7, then becomes inadequate to excite the resonant circuit in the sense of the preceding discussion. The pattern of operation of the switching matrices is then modified so that a succession of equivalent circuits occurs corresponding to the sequence shown in Fig. 10(d). The eventual purpose of creating a current pattern I_i is shown in Fig. 10(c). It means that the current

$$I_{s\ av} = -I_{i\ av} \tag{7}$$

leaves the converter and enters the source of electric energy with voltage e_s. The current $-I_{i\ av}$ of equation (7) is determined by the pattern of switch operation referred to previously and is independent of equation (6). Braking of a machine can be therefore implemented for any rotation speed of the machine and of the "motor" voltage V_0. Operation of the third and the fourth quadrant is analogous to the first and second quadrants described above.

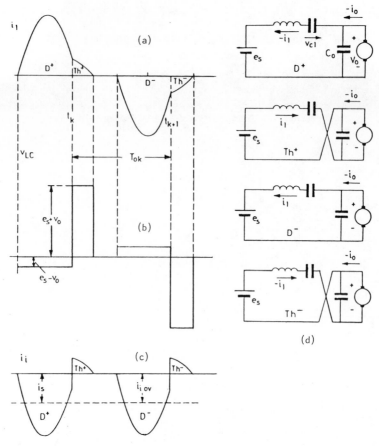

Fig. 10   Current and voltage waveforms for second-quadrant operation: (a) the resonant current I_1; (b) the driving voltage V_{LC}; (c) the input current I_i to
the converter and (d) the succession of four recurrently valid equivalent
converter circuits.

Fig. 11 illustrates the characteristic aspect of successive four-quadrant operation of
the converter discussed in schematic form.

The four-quadrant illustration shown in Fig. 9 is extended by indication of the current I_0 concerned and the associated motor voltage as functions of time. Superimposed
on the current I_0 is an indication of the sequence of current pulses $|I_1|$, with appro-
priate polarity, commensurate with the intended purpose of four-quadrant operation.

The time functions I_0 and V_0 start at the upper right corner (quadrant I) of the
four-quadrant operation. The origin is encircled, following counterclockwise the rec-

tangular path in the V_o - I_o plane, as indicated in Fig. 11. The motor current I_o is comprised of piecewise straight lines below the V_o - I_o plane and follows the four-quadrant operation presented. The synthesizing current pulses $|I_1|$ are idealized with equal repetition rate as forming the "integrands" of I_o during the dynamic phases of I_o and the "averages" during the phases of steady-state operation of I_o. The V_o "curve" can be interpreted in an analogous manner.

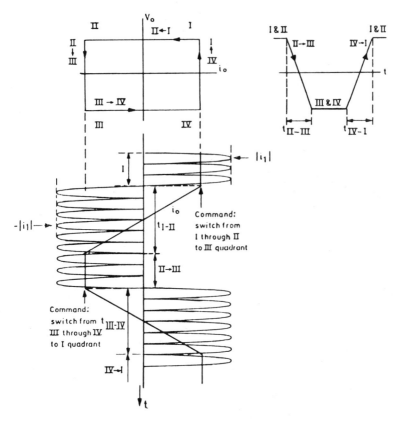

Fig. 11   Schematic presentation of motor currents and voltage associated with the four-quadrant operation of a d.c. machine.

One of the remarkable aspects of the process described above is the fact that the power-transferring output current $|I_1|$ can reverse its polarity without delay. This behavior is the result of unique system properties, which include a predicting output overcurrent- and an inherent short-circuit protection. Actual CRT traces of the processes which were idealized in Fig. 11 are shown in Fig. 12. The response of the motor is governed by the time constants involved.

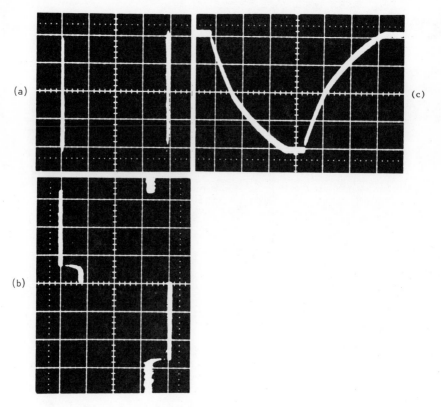

Fig. 12   CRT trace of (a) the input voltage V_o of the motor (100 V/div) as function of the motor current I_o (10 A/div); (b) the motor current $I_o(+)$ and (c) the motor voltage $V_o(+)$ ($t = 2$ s/div).

The maximum value of $|I_o|$ is limited by the relation

$$|I_o| \leq I_{max} \tag{8}$$

where I_{max} is a "precisely" preset maximum value of I_o and in itself limited by the thyristors applied. The reversal of polarity of the current form $+|I_1|$ to $-|I_1|$ is shown in Fig. 13. The recovery process for the polarity reversal of $\pm|I_1|$ can be clearly observed when examining Fig. 13. Fig. 14 shows V_o when the motor is "braked" at the end of its run or left to be halted by mechanical friction.

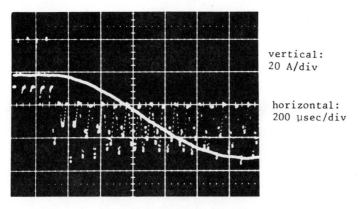

vertical:
20 A/div

horizontal:
200 µsec/div

Fig. 13   CRT traces of the selectively rectified resonant current $\pm|I_1|$ and of the motor current I_o.

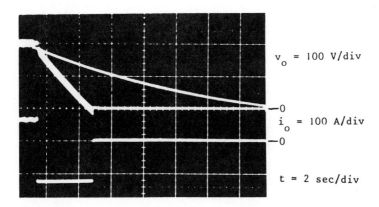

v_o = 100 V/div

—0

i_o = 100 A/div

—0

t = 2 sec/div

Fig. 14   CRT traces of the motor voltage V_o and the current I_o with and without application of the electric action of the converter.

A further point of interest is the motor current ripple of approximately 1 percent rms of the maximum d.c.-motor current, as illustrated in Fig. 15. The ripple of the motor current in Fig. 15 is also reproduced in a tenfold amplification of its actual current variations.

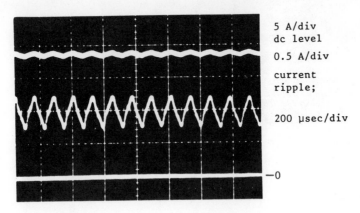

5 A/div
dc level

0.5 A/div

current
ripple;

200 μsec/div

Fig. 15   CRT trace of the motor current with its amplified ripple.

7. THE FOUR QUADRANT AC-TO-DC CONVERTER

A generalized form of the d.c.-converter circuit shown in Figs. 5 and 8 is presented in Fig. 16. The two-terminal input ports are enlarged to three terminals for connection to a three-phase network. The series resonant LC circuit performs the same function as described previously.

The output bridge, to the right of the schematic, performs an analogous function to the output bridge in Fig. 5. The switching matrix on the three-phase input side consists of twelve thyristors so that each of the six switch pairs can operate in complete symmetry with respect to the polarity of the connected a.c. phases of the three-phase supply line.

The system shown in Fig. 16 works, in principle, in the same way as those described previously, except in its capability to select two phases from which (or into which) it will work at any one time. This system can convert polyphase a.c. to d.c. without the interposition of low-frequency or d.c. filters. The a.c. ripple is removed by way of pulse modulation.

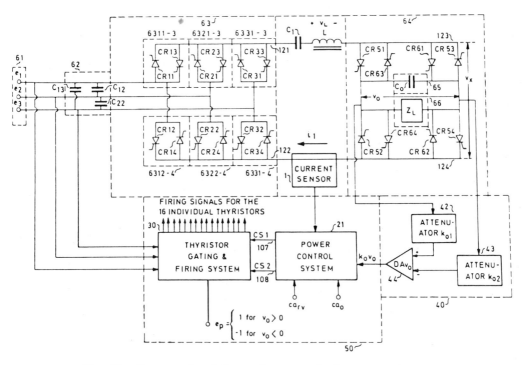

Fig. 16   Four-quadrant three-phase a.c.-to-d.c. converter.

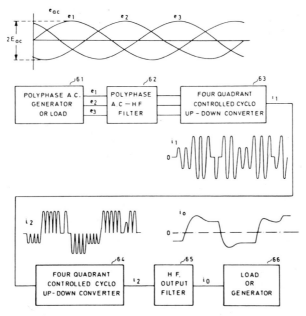

Fig. 17   Critical waveform of three-phase a.c. to single phase a.c. conversion non-sinusoidal).

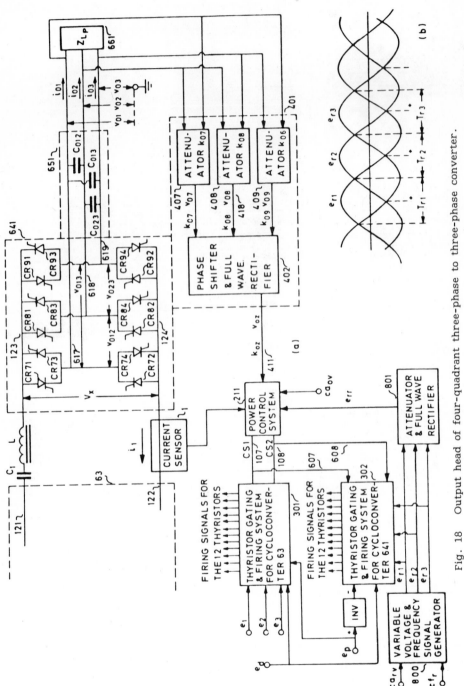

Fig. 18   Output head of four-quadrant three-phase to three-phase converter.

The process of waveform transformation from three-phase a.c. to single phase a.c. is indicated in Fig. 17. Generation of single phase nonsinusoidal a.c. was chosen for illustration of the capability of this system.

A three-phase output head is shown in Fig. 18 attached to the input port depicted in Fig. 16. The electronic control system is shown in the form of a block diagram. Certain aspects of the resulting voltage waveform transformation are indicated in Fig. 19. A set of lower frequency phase current waveforms I_{oj} (j = 1,2,3) is derived from a three-phase source with voltage e_m (m = 1,2,3) indicated in Fig. 19(f). The governing reference waveform is shown in Fig. 19(a).

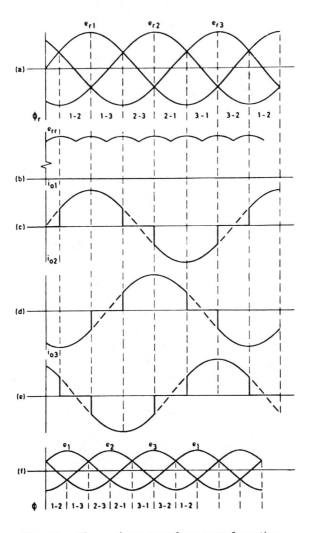

Fig. 19   Three-phase waveform transformation.

8. A FEASIBILITY MODEL OF AN AC-TO-DC CONVERTER

An experimental model of a three-phase a.c.-to-d.c. converter for 3 kW power was devised and tested. CRT traces of some of the significant waveforms are shown in Figs. 20-22.

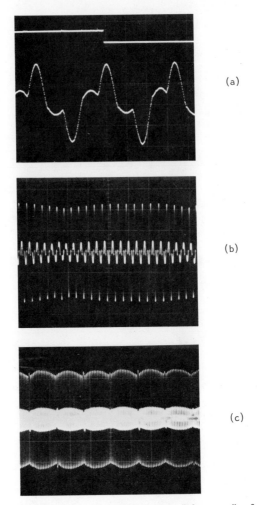

(a)

(b)

(c)

Fig. 20 CRT trace of the resonant current I_1: (a) "close-up" of the actual waveform; (b) one sixth of the 50-Hz cycle; (c) the resonant current during one 50-Hz cycle.

Fig. 20 illustrates the "resonant" current on various time scales. Fig. 20(a) shows a "close up" of the resonant current, in which the variations of the input voltage are not perceptible. One of the six intervals of the a.c. rectification process is discernible in the presentation of the resonant current trace of Fig. 20(b). A full cycle of the 50-Hz rectification process is depicted in Fig. 20(c).[3] Fig. 20 conveys a pictorial impression of the time resolution of the process discussed.

The processes of "forward" and "reverse" transfer of energy are illustrated in Fig. 21. The current I_i in the appropriate phase inside the converter is clearly seen in part (a) as being in phase with the associated phase voltage. The current pulses appear with the opposite polarity versus the phase voltage in part (b) of the same figure which illustrates "reverse" transfer of power.

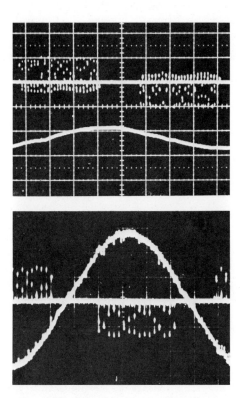

Fig. 21   Traces of the phase current I_i inside the converter and of the voltage e_i associated with one switch pair for (a) controlled a.c./d.c. conversion and (b) its reverse.

The CRT trace of the voltage across one of the six thyristor pairs is shown in Fig. 22, when viewed during one 50-Hz cycle. The necessity of applying fast-switching symmetrical thyristors becomes immediately apparent, as this pair works under equal external conditions in both directions of polarity during the cycle.

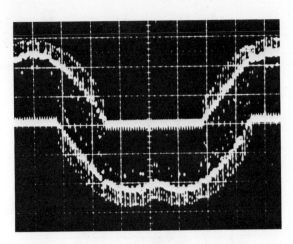

Fig. 22   Anode-to-cathode voltage of one thyristor of the three-phase bridge during one 50-Hz cycle.

REFERENCES

1. F.C. Schwarz and N.C. Voulgaris, "A Wide-Band Wattmeter for the Measurement and Analysis of Power Dissipation in Semiconductor Switching Devices," IEEE Trans. Electron Devices, ED-17, No. 9 (September 1970) 750-755.

2. F.C. Schwarz, "An Improved Method of Resonant Current Pulse Modulation for Power Converters," IEEE Trans. Industr. Electr. and Contr. Instr., IECI-23, No. 2 (May 1976) 133-141.

3. F.C. Schwarz and W.L. Moïze de Chateleux, "A Multikilowatt Polyphase AC/DC Converter with Reversible Power Flow and without Passive Low Frequency Filters," Proceedings 10th IEEE Power Electronics Specialists Conf. (PESC), San Diego, U.S.A., June 1979.

4. F.C. Schwarz and J.B. Klaassens, "A Controllable 45-kW Current Source for DC Machines," IEEE Trans. Industr. Applic., IA-15, No. 4 (July/August 1979) 437-444.

5. F.C. Schwarz and J.B. Klaassens, "A Reversible Smooth Current Source with Momentary Internal Response for Nondissipative Control of Multikilowatt DC Machines," IEEE Trans. Power App. and Systems, PAS-100, No. 6 (June 1981) 3008-3016.

6. F.C. Schwarz, "Analog Signal to Discrete Time Interval Converter (ASDTIC)," U.S. Patent No. 3,659,184, April 1972.

DISCUSSION
(Chairman: Prof. H.-Ch. Skudelny, RWTH, Aachen)

J. Vitins (BBC Brown, Boveri & Co. Ltd., Baden)
A major limitation of the system is the turn-off losses especially at high power levels.
If you use a device of 700 A or so that works well at 400 Hz you would burn it up
at 4 kHz. The only way to get around this point today is to use some kind of dI/dt
limiting chokes.

F. Schwarz (Delft University of Technology, Delft)
You have here a circuit in which the inductance has a necessary beneficial action in
contrast to the older so-called parallel-inverter circuits. If you want to turn-off a circuit with forced commutation then you want minimum inductance. You consider it as a
parasitic aspect of the circuit, but in this case the inductance is actually a necessary
component that has to be there anyway so it is no drawback to put it in, and use it
as a dI/dt-limiter.

H.-Ch. Skudelny
I am somewhat concerned about the efficiency figures you gave us because as I saw
there are at least four semiconductor devices connected in series. We are operating
at very high frequency with oscillating currents which also produce some losses in
the coils and capacitor so how do you come to such high efficiency figures?

F. Schwarz
The efficiency figures given were obtained at input voltages of the order of 600 V.
Another thing is that the capacitors, which were especially developed for this type
of converter by NASA, transfer between 5 and 10 kW, with a most impressive rate of
rise of temperature 8 to 10°C. Most European capacitors do not give you this type of
efficiency. These are not of a special construction: the entire trick is that the bands
are rerolled to remove the burr from the sides of the tapes, special machines are used
for winding the capacitors and, most important, is the use of well-suited contacts on
the side because these capacitors have been created in their communication type image.
Today you get good and heavy copper contacts of braids coming out that really can
carry the current and have the necessary joining techniques.

H.-Ch. Skudelny
What about the inductors?

F. Schwarz
Only two years ago we used inductors having molybdenum permalloy cores for that
purpose and we had losses that were about four times as high as calculated even
though we used stranded wire. We then built a 200-kW unit with inductors weighing

200 kg. Today we construct these inductors with air cores and with stranded wire. The inductors give us losses that are indistinguishable from the losses that would occur under d.c. power conditions at these frequencies. Today the weight of these inductors is about 37 kg with higher efficiency because there are no core losses at all. All high-frequency transformers can be constructed today with ferrite cores which makes them relatively cheap. Another point is that together with the air-core inductors the inverter has become unusually quiet because the only noise is from the thyristors themselves. There is a reduction of noise by an order of magnitude. And with 30 kHz or even 20 kHz inversion there is in any case no audible noise. Now we intend to install meters in our laboratories to protect personnel against noise that is above the audible level.

N. Coulthard (Thomson CSF, Courbevoie)

Would you agree that the use of gate-assisted turn-off thyristors would be well adapted to such a circuit to give you initially a very low turn-off time and also to apply to the dV/dt problems mentioned by Dr. Vitins?

F. Schwarz

The system was originally developed using just ordinary thyristors and one of the surprises was that the losses were considerably lower than expected from the manufacturer's tables. In the last one or two years, we have started to use gate-assisted turn-off thyristors. We use this for two purposes. First to shorten the turn-off time, as with the Siemens BST-49's we can go down to 6 μs and achieve a 20 kHz inversion frequency. The second reason for applying negative bias is to increase the dV/dt that can be applied. There is an advantage in the use of gate-assisted turn-off thyristors and the advancement of the thyristors has greatly improved the system as such.

P. Knapp (BBC Brown, Boveri & Co. Ltd., Baden)

You use high-quality RC series circuits which have the effect of multiplication of voltage so that even in low-power installations you may have to consider very high voltages due to the voltage multiplication of these resonances.

F. Schwarz

What you probably mean is that the capacitors would swing out of proportion with regard to the voltages. This was one of the crucial elements in developing this system and has plagued the system for many years. But about seven years ago, a philosophy was found that would contain all voltages and currents as part of the control mechanism and in this way it was possible to use these systems. This system has been tested by NASA for something between 20,000, and 30,000, hours in thermal vacuum with one short circuit applied to the output terminals every two minutes and one open circuit every two minutes in an asynchronous manner to prove the reliability of the system. They plan to use it to power their ion engines for space. And these ion engines sputter. At an output voltage of 1500 V the system has survived for four years.

REVERSE-CONDUCTING THYRISTORS[*]

P. DE BRUYNE, J. VITINS and R. SITTIG
BBC Brown, Boveri & Co. Ltd., Baden, Switzerland

SUMMARY

The application of semiconductor devices in power installations is continuously growing due to the improved performance and the resulting cost reductions of such systems. This advancement is made possible by adapting the device characteristics to the circuit requirements. This is particularly the case with reverse-conducting thyristors. In applications where soft commutation through an antiparallel feedback diode is used, new asymmetric device structures are possible which double the power-handling capability for a given turn-off time. By integrating the feedback diode into the asymmetric thyristor chip the dynamic devices properties are further improved and the number of power devices for a given circuit is reduced to a minimum. The analysis of important applications, e.g. PWM-inverter and chopper circuits, is used to demonstrate the ability of reverse-conducting thyristors to reduce the costs and the volume of installations, and to open new high-frequency applications. The important fabrication techniques of the reverse-conducting thyristor are described.

1. INTRODUCTION

Power semiconductor devices have the principal purpose of turning on and off the flow of electric power. In the range of low voltages, V < 1000 V, and at low and medium power, P < 10-15 kW, modern power transistors fulfill this function very well in many applications. In the range of higher powers up to several 100 MW, for example, in HVDC converters, self-injecting thyristors and diodes are more economical. Conventional thyristors are designed for a forward- and reverse-blocking capability. The optimum device characteristics are found in a compromise between switching time and blocking voltage. In soft commutation circuits the reverse-blocking capability is not required. This can be utilized to reduce the chip thickness and to introduce new asymmetric device structures. With these a new compromise is obtained. At a given blocking voltage the turn-off time can be reduced by a factor of three.

---

[*] Presented at the symposium by P. De Bruyne

The discrete asymmetric thyristor differs from the reverse-conducting thyristor (RCT) which has an integrated antiparallel diode. The monolithic integration improves the dynamic properties.

The RCT was first proposed by Kokosa and functional devices produced several years later by various companies.[2,3,4] With the improved electrical characteristics and the concomitant reduction of costs, weight and bulkiness, the RCT has contributed to the advancement of the state of the art in many power applications.

2. STATIC CHARACTERISTICS

Diodes and thyristors are self-injecting bipolar devices. In the on-state these devices are flooded with charge carriers which are injected from the highly doped n^+ and p^+ emitter diffusions. The injection increases with increasing current with only a small variation in the forward voltage drop. The exact value of the on-state voltage is determined by the charge-carrier lifetime, the total width between the n^+ and p^+ layers and the injection efficiency of these layers, and can be calculated accurately. Fig. 1

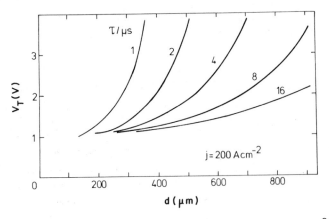

Fig. 1   Forward voltage drop V_T at a current density $J = 200$ A/cm^2 as a function of the silicon thickness, d, for various charge-carrier lifetimes τ. The turn-off time t_q of thyristors is typically $15.\tau$.

shows the results of such calculations with the assumption of a constant charge-carrier lifetime due to recombination centers and with an emitter efficiency of one.[5] The switching speed is roughly proportional to τ and the maximum blocking voltage is related to the device thickness. It is evident from Fig. 1, that for a given switching speed, the maximum device thickness is limited by a rapid increase of the forward voltage drop. The doping profile and the electric field distribution in the blocking state are shown for a conventional thyristor in Fig. 2a. The maximum blocking voltage

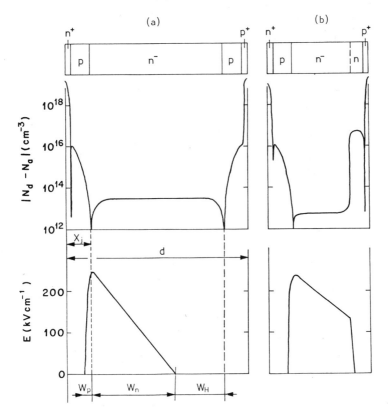

Fig. 2   Schematic doping profiles and electric-field distributions in a) conventional thyristors and b) asymmetric devices with an n-blocking layer. A significant reduction of silicon thickness, d, is possible with the asymmetric devices compared to conventional thyristors. X_j is the diffusion depth of the p-base, W_p and W_n are the space-charge regions in the p and n^--bases respectively and W_H is the neutral zone.

is obtained when the breakdown field E_m reaches 2.10^5 V/cm. The field gradients are determined by the doping profiles in the p and n^--base. The blocking voltage is given by the area of the field distribution. The device thickness must be increased by the neutral zone W_H. This region must be large enough to ensure the recombination of charge carriers injected from the forward-biased anode emitter. The blocking voltage is increased by lowering the base-doping concentration which makes a larger device thickness necessary. An inhomogeneous doping concentration must be taken into account by further increasing the silicon thickness. This can be avoided by using neutron transmutation doped starting material.[6]

When the reverse-blocking capability is not required, improved device structures are possible. A reduction of approximately 60 μm in silicon thickness is possible just by avoiding the deep p-diffusion in the anode. The device thickness can be greatly reduced further by introducing an n-blocking layer into the anode (see Fig. 2b). With this layer and with a low base-doping concentration, the average electric field can be greatly increased. The electric field at the n^- - n interface is often chosen to be $E = 0.5\ E_m$.[7]

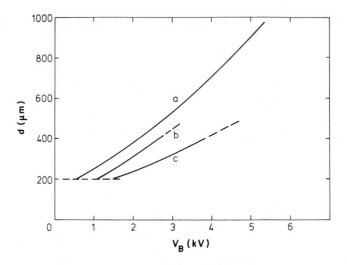

Fig. 3   Device thickness, d, versus breakdown voltage, V_B, for a) conventional thyristor structure, b) asymmetric device with minimum silicon thickness, however, without an n-blocking layer and c) asymmetric device with an n-blocking layer.

In Fig. 3 the required device thickness is shown versus the breakdown voltage for the above three structures. It can be seen that the voltage ratings can be increased by 80 % for devices with an n-blocking layer compared to conventional reverse-blocking thyristors of the same silicon thickness. A minimum device thickness is assumed at d_{min} = 200 μm due to handling limitations. In all cases a p-base thickness of 70 μm was employed.

With the minimum device thickness given above, the minimum turn-off time can be obtained from Fig. 1. The charge-carrier lifetime τ, and accordingly the turn-off time, $t_q \sim 15\ \tau$, is determined by the maximum allowable on-state voltage drop. For self-injecting devices this is limited by the available cooling systems and is here chosen to be $V_F = 1.5$ V at $J = 50$ A/cm^2 (corresponding to the mean current I_{TAVM}). The resulting turn-off times versus the blocking voltage are shown in Fig. 4, again for the three structures discussed above.

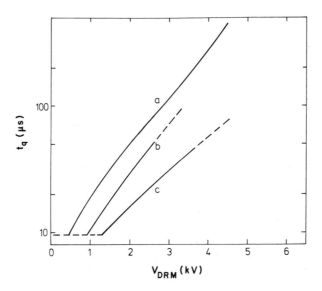

Fig. 4   Obtainable turn-off time t_q versus blocking-voltage rating, V_{DRM}, assuming a forward voltage drop $V_T^q = 1.5$ V at $J = 50$ A/cm^2. a) conventional thyristor structure, b) asymmetric structure without an n-blocking layer and c) asymmetric device with an n-blocking layer.

The substantial improvement in the switching characteristics obtained with the asymmetrical thyristors is evident. It can be seen that two conventional reverse-blocking thyristors with $V_{DRM} = 1500$ V and $t_q = 30$ μs can be replaced by a single asymmetric device with $V_{DRM} = 3000$ V. Both devices have the same specific power losses.

3. DYNAMIC CHARACTERISTICS

3.1 Turn-on

The turn-on process in a thyristor can be described in two steps.[8] The first conduction channels are formed at the emitter edge near the gate. The lateral extent of the first channel in the gate area is dependent on the gate current on the one hand, and on the doping profiles, the gate geometry and the masking precision on the other. The dI/dt capability at turn-on is given by the initial turn-on area. In the second step the ignited area spreads over the entire emitter. The spreading velocity can be of order 20 μm/μs. It depends primarily on the following: 1) the charge-carrier lifetime, 2) the lateral structure, and specifically the emitter-short design, and 3) the current density and therefore directly on the application. As only a part of the thyristor is ignited in the spreading phase, the on-state slope resistance will increase in inverse proportion to the ignited area. The power losses during turn-on are therefore larger than during the static on-state.

The increased power losses and temperatures at turn-on must be given special consideration with large devices, i.e. with silicon diameter greater then 10 mm, and at switching frequencies f > 100 Hz. The turn-on behavior can be greatly improved by increasing the area of the initial conduction channel. A further design objective is to distribute turn-on in the main thyristor with a minimum loss of active area without overloading the amplifying gate (see Fig. 5). The functioning of the fingergate structure can be observed with infrared recombination radiation as shown in Fig. 6.

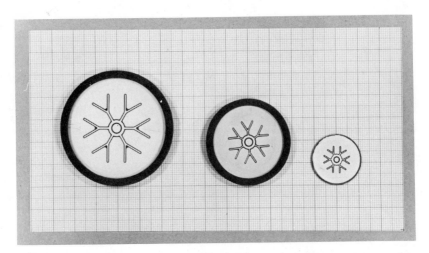

Fig. 5   Silicon chips of reverse-conducting thyristors. Form left to right: CSR 727 ff, CSR 327 ff and CSR 146 ff. The interdigitated fingergate structures reduce turn-on losses and temperature excursions. The thyristor part is in the center with the integrated diode in the outer rim.

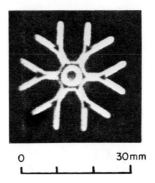

O   30mm

Fig. 6   Infrared recombination radiation at turn-on of the CSR 731-20. With the fingergate structure the initial turn-on area is distributed over the entire
emitter.

With the above analysis of the turn-on process a model framework can be developed
from which the power losses and junction temperatures can be calculated at any time
during the turn-on period.[8,9] To determine the maximum temperature, the variation
of the thermal resistance with the ignited area must be taken into account. This tool
is invaluable when designing fast-switching thyristors, especially RCTs, so that they
will perform optimally in the intended applications. Furthermore, the optimum utilization of existing devices in real circuit applications can be determined accurately. This
helps to reduce development costs for new circuits and increases long-term reliability.

In Fig. 7 the power losses of an RCT with a fingergate structure, the CSR 731-20ix,
is compared with an earlier design, the CSR 751-20ix,[4] which has a simple central amplifying gate. Both devices were chosen with a blocking voltage of V_{DRM} = 2000 V, t_q
(DIN) = 35 µs, and the same on-state characteristics in the diode and the thyristor.
The comparison is made with a damped sinewave current pulse of $\hat{I}_T$ = 2000 A, $\hat{I}_D$ =
1750 A and t_p = 150 µs. The large reduction in the thyristor losses with the fingergate structure is evident. The temperature ripple which is especially important in
pulse applications is also greatly reduced, typically by a factor of 2 - 4. The comparison between measurement and calculation is also shown in Fig. 7.

The turn-on behavior in typical high-pulse chopper or PWM-inverter applications is
shown in Fig. 8, for the same thyristor structures as above but now with V_{DRM} =
1600 V and t_q (DIN) = 30 µs. The comparison is made with a current waveform which
typically occurs in fast-switching power applications, e.g. in chopper circuits. Whereas the CSR 729 is almost fully turned on by the commutation pulse alone, the CSR 749
requires a t_p > 500 µs to be fully ignited at the end of the current pulse. The in-

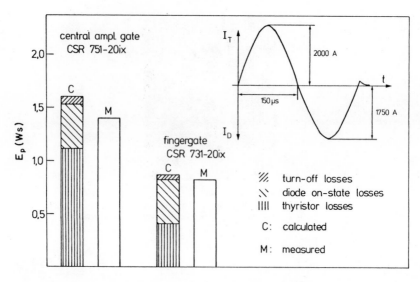

Fig. 7   Comparison of energy losses, E_p, between a simple amplifying gate structure, CSR 751-20ix, and a fingergate structure, CSR 731-20ix. Both devices
have the same on-state characteristics. Good agreement between calculation,
C, and measurement, M, is obtained.

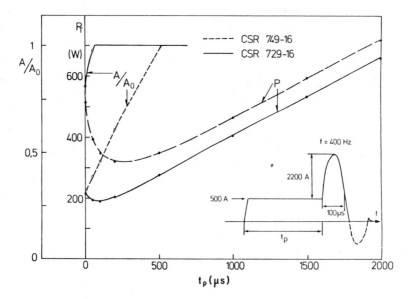

Fig. 8   Power losses, P_T, and relative turn-on area, A/A_0, of the CSR 749 without
fingergate and of the CSR 729 with fingergate. The comparison is made with
identical on-state characteristics. Two-inch silicon chips with simple amplifying gate structures can take more than 500 μs to be fully ignited. The current waveform is typical of fast-switching applications.

creased power losses in the CSR 749 at short t_p are due to the full commutation current passing through only a small ignited area of the device. Even at long pulsewidth the fingergate structure gives a marked reduction in the thyristor losses.

3.2 Turn-off

Both the discrete asymmetric thyristor and the RCT are turned off by soft commutation. In the ideal case, the reverse voltage is given by the forward voltage drop of an antiparallel diode. During this time the injected charge carriers in the thyristor recombine. The dynamics of the turn-off process, however, are different for the two devices.

The turn-off time t_q is principally determined by the rate of decline of the thyristor current, dI/dt, the reverse voltage during commutation, and the voltage rate of rise, dV/dt, to a given voltage overshoot, V_D. It is evident that the dynamic current and voltage waveforms greatly influence the turn-off time. Well defined turn-off conditions therefore simplify circuit design and increase the operating reliability. By integrating the feedback diode into the RCT, turn-off is improved compared to the discrete asymmetric device in the following ways.

With the RCT the reverse voltage during commutation always equals the forward voltage drop of the integrated diode and is therefore independent of stray inductances even at the highest allowable dI/dt.[10] Because a greater negative voltage can never be applied to the device, the measured turn-off time at defined laboratory conditions will be the same as in real circuit applications.

With the discrete asymmetric thyristor and diode the reverse voltage is influenced by stray inductances, L_σ, between the two components as shown in Fig. 9. The inductive voltage is comparable to the on-state voltage drop, e.g. for $L_\sigma > 0.01$ μH and dI/dt ∿ 100 A/μs. The stray inductance produces a commutation voltage which is very different from the defined laboratory conditions given in data sheets. The negative voltage can be increased with L_σ, however, with the disadvantage of a zero crossing and a high positive inductive voltage in the latter half of the commutation period. Because the reverse voltage does not clear the p-n junction of charge carriers for forward blocking, it is difficult to determine in advance the effect of the inductive voltage. Therefore, either additional safety margins must be employed or, for optimum design, experiments must be carried out under real circuit conditions.

The maximum reverse voltage of the asymmetric device is given by the breakdown voltage of the anode p-n junction. Due to the relatively high doping concentration of the n-blocking layer, this voltage is of the order of 20 V. It is evident, that the above inductive voltage can seriously limit the application of such a device in real circuits. Close proximity between the thyristor and the feedback diode must always be guaranteed.

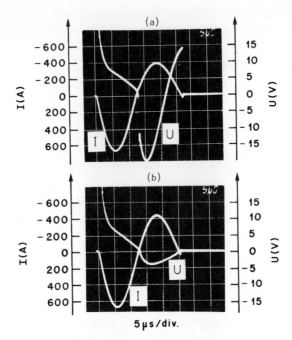

Fig. 9   Commutation voltage a) using a discrete feedback diode and b) with the RCT. In a) the commutation voltage waveform depends on the mechanical design and on dI/dt.

The voltage rate of rise dV/dt, at turn-off is dependent on the reverse recovery current waveform of the antiparallel diode together with the applied snubber circuit. With the discrete asymmetric thyristor, the dV/dt ratings of the thyristor, the recovery characteristics of the diode and the snubber components must all be chosen to match. With the RCT circuit design is simplified, as the thyristor and diode parts are designed, produced and tested together. Because it is processed as a single device, the dV/dt capability and the reverse recovery of the diode always match. Hence, a device with t_q (DIN) = 4 μs can typically have an inherent dV/dt capability in excess of 2000 V/μs and the reverse recovery waveform is permitted to be sharper, leading to off losses. Such a device is ideal for high-frequency applications. On the other hand, with t_q (DIN) = 40 µs, the inherent dV/dt capability can be greater than 1000 V/µs.

The matched reverse recovery of the diode and the absence of stray inductances makes an accurate calculation of the commutation circuit possible with the RCT. The dependence of t_q on dI/dt, dV/dt and V_D can be given for each RCT.[11] For a given commutation current pulsewidth it is often possible to employ a longer t_q (in practice up to 40 %) compared to a discrete solution. This gives a higher power-handling capability. The other possibility is to reduce the commutation current pulsewidth with the RCT.

3.3 The dI/dt and dV/dt Capability

The dI/dt and dV/dt capability of both devices is similar to that of conventional thyristors. With an appropriate design of the gate structure a dI/dt rating at turn-on above 400 A/µs followed by a full-load current is possible.

In real applications the dI/dt capability is not limited by thyristor turn-on, but is determined by the reverse recovery characteristics of the discrete or integrated feedback diode. At high dI/dt the reverse recovery current is large and the current recovery waveform generally becomes sharp. The maximum dI/dt is given by the resulting dV/dt and voltage overshoot which should not exceed the maximum guaranteed ratings of the thyristor. It must be controlled by the circuit design and by the snubber components.

The dV/dt capability of the device is adjusted by the emitter-short structure. Too heavy shorting, however, hampers the spreading of the ignited area. In contrast to phase-control thyristors, the emitter shorts in asymmetric devices are also designed to reduce the turn-off time by preventing injection by the charge carriers which still remain from the previous load current. Of course, the injection regions in the thyristor and the diode of the RCT must be perfectly insulated. If this is not achieved charge carriers from the flooded diode part can diffuse into the thyristor and cause turn-off failure. For fast-switching devices a distinction must be made between the static and dynamic dV/dt capabilities. Generally, the static dV/dt is several kV/µs while the dynamic ratings are determined by the turn-off conditions and are often defined with dV/dt = 1000 V/µs to V_D = 0.67 V_{DRM}.

4. DESIGN AND RELIABILITY

A cross section of the RCT is shown schematically in Fig. 10. A one-sided aluminum diffusion forms the blocking p-n junction which extends through the thyristor and the

diode of the RCT. In the devices shown in Fig. 5, the thyristor with its interdigitated fingergate structure is located in the center and the diode is placed in a concen-
tric ring around it.

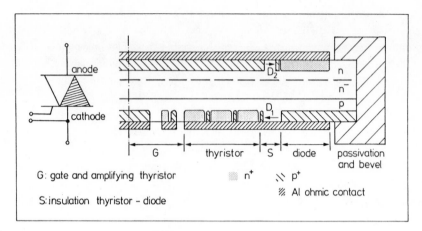

Fig. 10   Schematic cross section of an RCT.

Fig. 10 shows n^+ and p^+ surface diffusions in both the anode and the cathode. These
are necessary for a high-injection capability. Aluminum layers are used as ohmic contacts. The free-floating silicon chip is mounted in a direct pressure contact between
two molybdenum discs.[12] This technique is well established for silicon diameters up to
100 mm.

Because the silicon chip is not alloyed, the surface doping in the insulation region can
be chosen as required. The p^+ diffusion ring, D_2, in the anode (Fig. 10) reduces injection into the insulation zone, while in the cathode, D_1, it functions as an annular
short surrounding the emitter. With the above design, the dV/dt capability in the thyristor is totally independent of the current amplitude in the diode, even for long
turn-off times.

The n-blocking layer can be obtained by a phosphorus predeposition or by ion-implan-
tation followed by deep diffusion or by epitaxial growth.[13] The blocking layer is generally 30 - 50 µm deep. The doping concentration is determined by the electric-field
distribution and is typically of order 10^{16} cm^{-3}. The dimensioning of the neutral zone
differs from conventional thyristors as drift fields and local variations in the charge-
carrier lifetime are important.

The occurrence of a high electric field at the n^--n interface makes an appropriate
surface bevel necessary. A positive bevel of about 45° is convenient and well adapted
for the main p-n junction. For the n^--n interface, however, this would represent a

negative bevel which could cause the electric field to reach values even higher than at the p-n junction. The application of a second bevel of about 3° is sufficient to reduce the field maximum at the interface.

The long-term reliability of power semiconductor devices is primarily determined by two characteristics: 1) the thermal cycling capability, and 2) the stability of the blocking characteristics. During thermal cycling the various contact materials within the housing and in contact with the heat sink rub and gradually wear. This can eventually bring about a higher forward voltage drop (higher power losses) and degrade the thermal resistance leading finally to a device failure. As most power devices are relatively new and operating experience of order 10 - 20 years is scarce, a judgement of the thermal cycling capability can only be obtained from laboratory testing and from an analysis of the contact materials and their mechanical tolerances. Using appropriate materials the direct pressure contact permits a thermal cycling capability in excess of 10^5 cycles at a temperature excursion $\Delta T = 100$ K. This value is obtained only with stringent mechanical tolerances, e.g. in surface flatness and roughness, of all contact partners.

The long-term stability of the blocking characteristic is given by the processing technique, the passivation and the housing design, e.g. hermetically sealed construction. High stability is reached with a double-passivation technique developed for high-voltage diodes, $V_{DRM} = 5$ kV. The passivation is controlled regularly with 1000-hour tests. The airtightness of each device is checked individually.

Because the RCT is a single device with one blocking p-n junction, one bevel and passivation and is packaged in the same way as other discrete components, the reliability of the RCT corresponds to conventional devices with similar technology. The reliability of the total circuit is enhanced as the total number of power components is reduced. This is shown in the following application examples.

5. APPLICATION EXAMPLES

Depending on the ratio of the thyristor to diode current ratings, two classes of RCTs must be distinguished. In the first class diode area is small and is designed for commutation purposes only. It must carry the commutation current pulse and the turn-off losses. This leaves the larger part of the chip for a maximum thyristor current rating. An RCT in a three-inch housing can replace a discrete thyristor and diode each in two-inch housings, and in many cases, due to its better utilization of the silicon area, even a three-inch thyristor and a two-inch diode. The CSR 727 ff, for example (see Fig. 5), has a current ratio of about 3:1 and is largely employed in chopper circuits for traction applications.

In the second class of RCTs the diode part is designed to carry the full-load current. Here the current ratio is 1:1. These devices are intended for PWM variable-speed motor drives with regenerative braking. In other voltage source applications, for example, uninterruptable power supplies (UPS), the mean diode current can be smaller than the mean thyristor current. Here, the diode power ratings are not fully utilized. Nonetheless, the overall improvement in circuit design and performance is still appreciable compared to the solution with discrete components. Because the combined power losses of the diode and the thyristor are dissipated over the same heat sink, it is possible to increase the load current in one part if that of the other part is reduced. In the extreme case where there is no current in the diode part, the power losses in the thyristor can be increased by more than 50 %, depending on the thermal resistance of the heat sink. The CSR 146 ff and the CSR 327 ff shown in Fig. 5, belong to this class of RCTs.

The design of the CSR 146 ff and the CSR 327 ff also permit optimal application in pulse circuits, for example, as commutation thyristors or in new electrostatic precipitators.

5.1 Calculation of Junction Temperature and Power Losses

With the RCT the maximum temperatures in the thyristor and the diode must be calculated separately. The thermal diagram of an RCT in a typical press-pack housing with heat sink is shown in Fig. 11. In the housing it is assumed that the heat flow is purely axial. This is generally a worst-case assumption. The lateral heat flow is generally small and depends on the power losses in the adjacent part of the RCT. For simplicity, the contact thermal resistance, R_{CK}, between the case and the heat sink is generally added to the thermal resistance of the heat sink. The power losses, P_T, in the thyristor part are composed of turn-on and static on-state losses. In the diode part the power losses, P_D, are made up of forward on-state losses and turn-off losses.

5.2 DC-Chopper

The d.c. chopper is commonly used in traction applications for variable-speed d.c. motor drives. With overhead-line voltages of nominal 600 V in short-haul traction up to 3000 V in railway applications, an optimal power circuit is obtained with high-voltage, high-power, fast-switching devices. The high-power ratings of the RCT combined with its improved switching capability have made possible great improvements in traction circuit design. In short-haul traction, for example, a d.c. chopper for P > 300 kW can be realized with only two RCTs: a main thyristor in a three-inch housing and a commutation RCT in a two-inch housing. This is a reduction in the number of power components by a factor of four compared to earlier designs using discrete devices with $V_{DRM} \leq 1400$ V. A typical application is shown in Fig. 12.

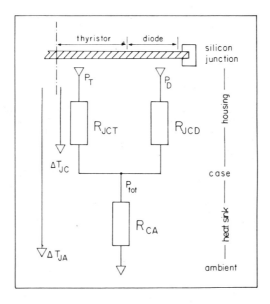

Fig. 11   Heat-flow diagram and thermal resistances of the RCT. Between the junction and the case heat flow is assumed to be strictly axial.

A chopper circuit with RCTs is shown in Fig. 13. The CSR 731-20ix (V_{DRM} = 2000 V) is designed as the main thyristor (TM) and is commutated with the CSR 331-20iz as TC. Commutation is accomplished by a full sinewave current pulse in the ringing circuit C_c, L_c, TC and TM. For high-voltage applications a series connection of several RCT's in the position of TM and TC is commonly employed.

The snubber circuit must be designed corresponding to the reverse-recovery characteristics of the diode part. The high-voltage RCTs generally have a very soft reverse-recovery current. With the indicated decoupled snubber (Fig. 13) the voltage ratings at turn-off and the inrush current at turn-on can be kept within guaranteed ratings. An overvoltage protection is possible with breakover diodes (BOD).[14] With BOD the current loading of C_c can also be limited.

Fig. 12   The RCT makes great improvements possible in the power circuits of trolley
busses, subways, trams, and high-power locomotives.

Apart from minimizing the number of power devices, the short turn-off times of the
RCT permit shorter commutation pulses. Thereby the current loading of C_c is smaller
and the passive components C_c and L_c can be reduced in size and weight.

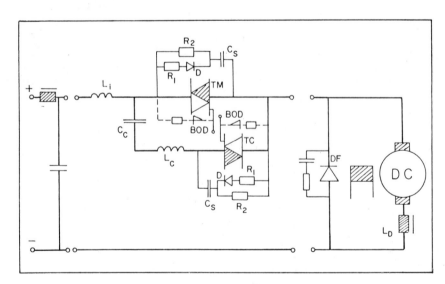

Fig. 13   Typical chopper circuit with RCT's. A decoupled snubber is generally recommended. Overvoltage protection is possible with the BOD. The chopper
and the free-wheeling diode can be mounted on a common heat sink.

In some cases it can be advantageous to employ very steep current pulses, dI/dt > 130 A/μs. Here, the snubber circuit must be chosen appropriately and in some cases saturable reactors are necessary. The CSR 731 and CSR 331 are designed for commutation pulses of typically Î > 3000 A, t_p ∿ 100 μs and f = 500 Hz.

In Fig. 14 the temperature excursions in the thyristor and diode part of the CSR 731-20ix are shown for a typical load cycle such as could occur in short-haul traction. For simplicity, the load and the commutation currents are kept constant throughout the full-load cycle. The load cycle consists of 1) acceleration with a linearly increasing duty cycle, $\alpha = t_{on}/T$, 2) driving with I (load) = 500 A and $\alpha = 0.98$, 3) braking with linearly increasing α, and 4) a stop. It is seen that the thyristor and diode temperatures follow each other closely. This indicates a perfect adaptation of the device to this application. For f = 400 Hz the total power losses in TM are $\bar{P}_T$ (tot) = 612 W max.

For comparison, the temperatures and power losses are also calculated for discrete components. The CSR 731-20ix is replaced by the CSF 551-20ix and the DSD 635-20.

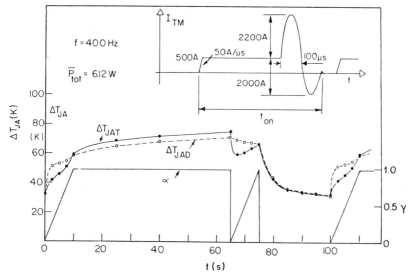

Fig. 14   Temperature excursions, ΔT_{JA}, in the thyristor and the diode of the CSR 731-20ix in a continous load cycle. The motor and commutation currents, I_M = 500 A and Î = 2200 A, are kept constant throughout the full-load cycle. The duty cycle, $\alpha = t_{on}/T$, is varied to simulate acceleration, driving ($\alpha = 0.98$) and braking. Thermal resistances: R_{CK} = 0.005 K/W and R_{KA} = 0.065 K/W.

The same commutation pulse was employed as above. The total power losses are now $\bar{P}_{tot} = \bar{P}_T + \bar{P}_D = 571$ W + 152 W = 723 W. Because the diode is operated at only a fraction of its power rating, it contributes unnecessarily to the total weight and space of the chopper.

5.3 Inverter and High-Frequency Applications

Due to its short turn-off times and the improved commutation conditions the RCT lends itself ideally to high-frequency applications.[11,15,16] In Fig. 15 the power module of a 4 kHz, 50 kW d.c. chopper is shown. The module includes the free-wheeling diode DF (Fig. 13). A very short commutation pulse of $t_p = 12$ µs and $\hat{I} \sim 400$ A is made possible with the super-fast CSR 147-10idl as TM and TC. These devices have turn-off times of typically $t_q \sim 4 - 5$ µs (DIN standard) or $t_{qs} = 10$ µs with dI/dt = -100 A/µs, dV/dt = 500 V/µs and $V_D = 800$ V, $T_{VJ} = 125°$C.

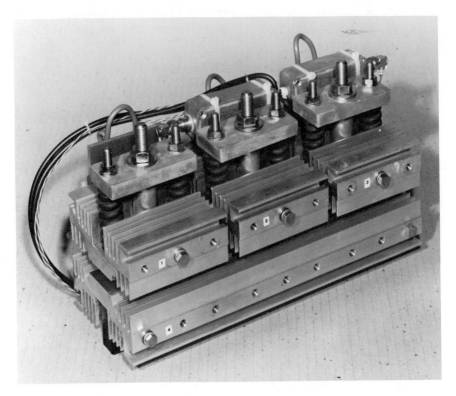

Fig. 15   Power module of a 4-kHz, 50-kW chopper with the CSR 147-10idl. A commutation pulse $t_p = 12$ µs and $\hat{I} \sim 400$ A is employed.

With the chopper just described a high power factor, which is practically independent of the load-current level and a low mains pollution, is achieved. The control circuit is simple, as the power is regulated independent of the input phase and with only two thyristors. The high operating frequency permits a reduction of the smoothing reactance L_D (Fig. 13). In some applications the motor inductance is sufficient, so that an external L_D is unnecessary.

Due to the fingergate structure of the CSR 147-10idl the turn-on losses do not increase substantially below about 4 kHz. The maximum switching frequency is given by the turn-off and snubber losses. With the CSR 147-10idl the frequency given above is possible with I (load) = I_{TAVM} and at maximum duty cycle α.

High operating frequencies are often desirable with PWM inverters to lower the harmonic content of the output waveforms and to simplify the filter design. Furthermore, short and high commutation pulses are needed to increase the modulation depth. The RCTs meet these requirements with their short turn-off times and their reduced turn-on losses. In Fig. 16 four basic circuits for voltage-source inverters are shown.   The

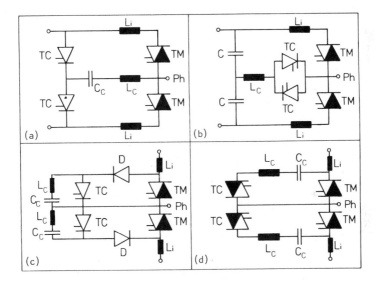

Fig. 16   Basic circuit diagrams for voltage-source inverters, e.g. PWM, using RCTs. TM is the main thyristor and TC is the commutation thyristor. Turn-off is accomplished with a current pulse in the ringing circuits C_c - L_c.

RCT as TM permits a smaller pulsewidth and a strong reduction in the number of power components. In a) and in some cases in c) the minimum pulsewidth is also determined by TC. In b) and d) the minimum pulsewidth can be very narrow as it is limited primarily by the turn-off time of the RCT as TM. The various circuits differ in the necessary commutation sequence and the various mechanical design possibilities. In b) and d) a full phase can be mounted on a common heat sink.

The temperature excursions ΔT_{JCT} of the CSR 147-10id are shown in Fig. 17 for a double-edge modulation with $\hat{I}$ (load) = 245 A, f = 50 Hz and $\hat{I}_c$ = 590 A, t_p = 22 µs. A 15- or a 21-pulse modulation will increase the total power losses by only about 15 and 25 %, respectively. The maximum calculated power losses are $\bar{P}_{tot}$ = 150 W in the 9-pulse application mentioned above. Higher commutation pulses at the same pulsewidth

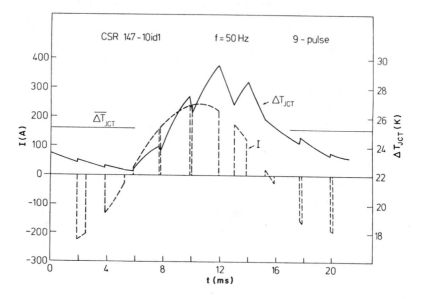

Fig. 17   Temperature excursion, ΔT_{JCT}, of the CSR 147-10idl in a 9-pulse double-edge modulated PWM application. Total power losses in the thyristor and the diode: $\bar{P}_{tot}$ = 150 W (max). Output current: $\hat{I}$ = 245 A, f = 50 Hz.

are possible. However, the snubber components must be chosen accordingly. Generally, the super-fast RCTs have a sharper reverse recovery and require a smaller snubber resistance R_1 (Fig. 13) than the high-voltage devices.

5.4 Pulse Applications

The high-pulse capability of the RCT permits its application as a turn-off thyristor. A strongly interdigitated gate structure is very important as the turn-on losses and the temperature ripple are greatly reduced permitting higher current pulses. Commutation is achieved with a single device in a ringing LC-circuit (Figs. 13 and 16). In the same way, high-current pulses for new electrostatic precipitators and for laser applications can be produced with a minimum amount of circuitry (Fig. 18). For high-voltage applications a string of RCTs in series connection can be employed. The number of RCTs in series is limited only by the static and dynamic voltage distribution. This can be well controlled by appropriate snubbers and for very large strings, $N \geq 20$, by surge-voltage suppressors, DSAS.[14] Generally the number of devices put in series is reduced by the high-voltage ratings of the RCT.

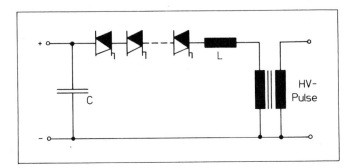

Fig. 18 Schematic circuit diagram for high-voltage pulse applications, e.g. electrostatic precipitators.

Typical current and voltage waveforms of the CSR 331-20ix are shown in Fig. 19 with $\hat{I}_T = 2300$ A, $\hat{I}_D = 2000$ A, $t_p = 56$ µs, and the maximum operating temperature, $T_{VJ} = 125°C$. The initial capacitor voltage was V = 1200 V. The snubber components $R_1 = 5.6$ Ω, $R_2 = 47$ Ω and $C_s = 0.5$ µF, were employed. The current and voltage waveforms at turn-off are also shown. The soft reverse-recovery-current waveform is typical of these devices.

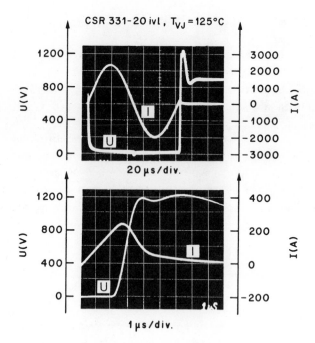

Fig. 19   Current and voltage waveforms with the CSR 331-20ivl employed in a high-power pulse application. a) $\hat{I}_T$ = 2300 A, $\hat{I}_D$ = 2000 A, t_p = 56 µs b) Turn-off waveforms with R_1 = 5.6 Ω, R_2 = 47 Ω and C_s = 0.5 µF; T_{VJ} = 125°C.

6. CONCLUSIONS

With asymmetric devices it is possible to double the power-handling capability of thyristors compared to conventional fast-switching devices. This can be accomplished by improved on-state characteristics or by higher voltage ratings, V_{DRM} > 3000 V. The dynamic operating characteristics are further improved with the reverse-conducting thyristor. The integration of the feedback diode eliminates problems imposed by the limited reverse-blocking voltage and simplifies circuit design. Furthermore, the number of power devices can often be reduced by 30 - 50 %. With the RCT power circuits can be made more cost effective and reliable.

REFERENCES

1. R.A. Kokosa and B.R. Tuft, "A High-Voltage, High-Temperature Reverse Conducting Thyristor," IEEE Trans. Electron Devices, ED-17 (1970) 667-672.

2. T. Yatsuo, T. Ogawa, Y. Terasawa, K. Morita, and K. Wajima, "A Diode Integrated High-Speed Thyristor," Proc. 2nd Conf. Solid State Devices, Tokyo, 1970.

3. H. Gamo, S. Funakawa, and J. Shimizu, "The Present Status and Applications of Power Reverse Conducting Thyristors," PESC, 77, 50-60.

4. P. De Bruyne, A.A. Jaecklin, and T. Vlasak, "The Reverse Conducting Thyristor and Its Applications," Brown Boveri Review, 66, No. 1 (1979) 5-10.

5. J. Cornu and M. Lietz, "Numerical Investigation of the Thyristor Forward Characteristic," IEEE Trans. Electron Devices, ED-19 (1972) 975-981.

6. M.J. Hill, P.M. van Iseghem, and W. Zimmermann, "Preparation and Application of Neutron Transmutation Doped Silicon for Power Devices Research," IEEE Trans. Electron Devices, ED-23 (1976) 809-813.

7. R. Kumar, D.J. Roulston, and S.G. Chamberlain, "Two-Dimensional Simulation of a High Voltage p-i-n Diode with Overhanging Metallization," IEEE Trans. Electron Devices, ED-28 (1981) 534-540.

8. A.A. Jaecklin and H. Lawatsch, "On the Dynamic Phase of Thyristor Turn-On," BBC publication CH-E 4.0309.OE and Bulletin SEV/VSE, Switzerland, 68 (1977) 299-303.

9. A.A. Jaecklin and H. Lawatsch, "A High-Speed Thyristor with Optimum Turn-On Behaviour," Brown Boveri Review, 66, No. 1 (1979) 11-16.

10. J. Vitins and P. Wetzel, "Rückwärtsleitende Thyristoren für die Leistungselektronik," BBC-Nachrichten, No. 2 (1981) 74-82.

11. J. Vitins and P. De Bruyne, "The CSR 146: A New 25-kHz Power Thyristor, "IEEE-IAS Conf. Proc. (1980) 695.

12. S.D. Prough and J. Knobloch, "Solderless Construction of Large Diameter Silicon Power Devices," IEEE-IAS Conf. Proc. (1977) 817-821.

13. P.M. van Iseghem, "p-i-n Epitaxial Structures for High Power Devices,"IEEE Trans. Electron Devices, ED-23 (1976) 823-825.

14. P. De Bruyne and P. Wetzel, "Improved Overvoltage Protection in Power Electronics Using Active Protection Devices," Electric Power Appl., 2.1 (1979) 29-36.

15. J. Vitins, "New Reverse Conducting Thyristors for Modern ac and dc Motor Drives," 3rd. Int. Power-Conversion Conference, Munich, 1981.

16. J. Vitins and O. Kolb, "Der rückwärtsleitende Thyristor, ein neues Bauelement der Leistungselektronik," Elektroniker, No. 5, EL 22 - EL 26 (1981).

DISCUSSION
(Chairman: Prof. H.-Ch. Skudelny, RWTH, Aachen)

N. Coulthard (Thomson CSF, Courbevoie)

You seem to indicate that the turn-off time of the asymmetrical thyristor with an external antiparallel diode and stray inductance was defined simply as the point at which the voltage reversed bias changed over. A lot of work by Campbell and Erickson, by General Electric, and ourselves has shown in fact that t_q is not simply defined by the time of the voltage changeover due to the cosine swing.

P. de Bruyne (BBC Brown, Boveri & Co. Ltd., Baden)

There are two possible switching points which could limit t_q of an asymmetrical thyristor. Neglecting first the voltage waveform during the induction time then one occurs with the high dV/dt at the end of the induction time. Due to the stray inductance, however, there is a voltage swing and a positive dV/dt at about half the induction time. This represents the second possible switching point. Using symmetric thyristors one can take advantage of the strong decrease of turn-off time on reverse voltage and even introduce an inductance between thyristor and diode. This is, however, not possible for asymmetrical thyristors. Therefore it is very difficult to ensure that they will not turn on at the latter switching point.

N. Coulthard

You wouldn't try to increase the reverse voltage. The fact that the voltage goes positive is not necessarily a bad thing because it will give you forward polarization and lead to a forward recovery mechanism which can actually improve the t_q.

D. Silber (AEG-Telefunken, Frankfurt)

Which type of edge contour bevelling do you have in the device?

P. de Bruyne

For the main p-n junction the best is to have a positive bevel of about 45°, but this represents a negative bevel at the n^+-n^- interface. To reduce the surface field there a new bevel of 3° is introduced.

P. Zimmermann (Robert Bosch GmbH, Erbach)

What are the possible operating frequencies of these devices?

P. de Bruyne

The operating frequencies are primarily given by the turn-on time and the time to ignite the total cathode area and secondarily by the turn-off time. Small devices such as the series CSR-147, we have operated up to 50 kHz in the laboratory. The total area is activated after about 15 µs and the turn-off time under DIN standards is 4-5 μs. For a realistic circuit at about 100 A/μs 500 V/μs this corresponds to a t_q less than 10 μs.

P. Zimmermann
Can you handle the full power under these frequencies?

P. de Bruyne
Yes, at 20 kHz we had a sinewave current with an amplitude of about 200 A.

P. Zimmermann
What would be the rating at 400 Hz? Would the current rating again be 200 A or more?

P. de Bruyne
It's about 300 A at 400 Hz.

OPEN DISCUSSION
(Chairman: Prof. H.-Ch. Skudelny, RWTH, Aachen)

H. Becke (Bell Laboratories, Murray Hill)
For optimum performance and safe turn-off, do you think it is sufficient to use a simple interdigitated four-layer structure with a gated p-base, sheet resistance as low
as possible to avoid the saturation phenomena (a) as shown in the figure, and (b),
with an n-base which is as wide as possible such that the on-state voltage, V_T, does
not become so high as to cause power dissipation problems.

M. Kurata (Toshiba R&D Center, Kawasaki)
You are exactly correct.

H. Becke
Can further benefits be derived by consideration of the lifetime control vertically in
the device such that the τ in the p-base is kept high while the τ_n in the vicinity of
the anode for n-p-n-p structure is kept very low?

M. Kurata
That's a very good idea. I actually talked about the increase in p-base lifetime while
keeping that in the n-base low.

H. Becke
Do you think you can derive further benefits from anode shorts for the purpose of
reduction of the recombination tail which gives rise to extra dissipation? Using shorts
you can extract the carriers more efficiently in the residual time.

M. Kurata
That sounds right in principle, but I don't really know if it's correct. The most recent products listed by Hitachi do have the anode short structure but they also have
a fairly large tail current, so that I'm not entirely convinced that introducing anode
structure will reduce the tail current.

H. Becke
My own experience is that to achieve faster response or higher frequencies with the
introduction of anode shorts you need less gold doping which is beneficial in other
ways. The disadvantage that the turn-on sensitivity is much reduced can be remedied
by other means. You can use Schottky-barrier diodes as indicated in your paper.
Then you have a nonregenerative region which helps the turn-off by avoiding the
bypass current through the bypass transistor in the turn-on so as to balance these

177

two characteristics. The third thing I wanted to ask is would it be beneficial to use a resistive cathode such that the plasma density would be restricted at high current densities?

M. Kurata
I think so.

F. Schwarz (Delft University of Technology, Delft)
What effect would a condition of natural commutation have on the turn-on and turn-off characteristics of the GTO?

M. Kurata
I don't have much experience in this area, but in principle it sounds as though in such a case you wouldn't need a GTO at all.

F. Schwarz
At first sight this may not be the case, but if it would help in the design considerations for the device to improve the turn-on or the turn-off characteristics then it could be significant. It would be another application for the GTO.

H.-Ch. Skudelny
Which would have the higher power efficiency, a converter consisting of GTOs or a converter consisting of thyristors and commutation equipment?

M. Kurata
I mentioned GTO converters are better in application. I don't think you have the possibility of improving efficiency in normal thyristor systems.

R. Jötten (TU, Darmstadt)
I have a question to the talk of Mr. Schwarz. Isn't his scheme similar to the Gigi scheme, the difference being that Gigi uses a parallel resonance circuit keeping it at approximately the same frequency and voltage. In that case control turned out to be possible but rather difficult because in control the voltage of this intermediate circuit is determined by the power balance between input and output and uses natural commutation on both sides. What surprises me with your system is the ease of control that it seems to have, whilst you have a series rather than a parallel resonance and the thyristor problems seem to be the same.

F. Schwarz
One of the primary considerations of using a series resonance circuit was to ensure that there is no d.c. connection between the source and the load, which is not guaranteed in a parallel resonant circuit. From the point of view of control it may become less easy to achieve control because in the case of a series resonant capacitor all that

is necessary is to measure the current through that capacitor and this is the only path, while in the other case again a situation of uncertainty may develop. The method of control of the system as a whole, and the protection of the capacitor and of the system against overcurrent and overvoltages, would not be as clear cut in a parallel circuit as it is in a series circuit.

H.-Ch. Skudelny
I should like to put a question to Mr. de Bruyne. You mentioned reverse-conducting thyristors and showed devices with diameters of two inches. Are larger devices now available or is it possible to build bigger devices?

P. de Bruyne (BBC Brown, Boveri & Co. Ltd., Baden)
Mr. Jaecklin showed a thyristor of 100-mm diameter, and I don't see any technical problems in making reverse-conducting thyristors of the same size. The problem is the market. Traction applications today are limited to trams, trolley-buses and locomotives at current ratings of about 800 A and these are marginal: normal applications are for about 600 A motor current. The next step can only be with much higher powers for inverters, for example for high-power locomotives, or frequency changers. I believe these new applications will come in the next few years and we shall have to develop reverse-conducting thyristors 3 or 4 inches in diameter.

R. Sittig (BBC Brown, Boveri & Co. Ltd., Baden)
Prof. Skudelny mentioned that it might appear that the three talks in this session are completely disparate and I would like to indicate why I chose to put them together. At present device designers have to decide between developing either turn-off devices, the GTOs, or devices with a short turn-off time, the reverse-conducting thyristors. This morning Dr. Stemmler stressed his preference for GTOs whilst Prof. Schwarz demonstrated that natural commutation represents a very interesting alternative. Therefore I would like to hear the comments of circuit designers concerning these two possibilities. What will be the more advantageous way to go - the short turn-off time or the turn-off capability?

A. Marek (Brown Boveri Research Center, Baden)
The user of these devices would like to achieve full protection against either a short-circuit overload or an open load circuit. We have heard today two extremes. In the first lecture I understood Mr. Stemmler to say that it is difficult to protect tradition-al thyristor circuitry against short circuit. On the opposite side, Prof. Schwarz claims that even with conventional components his circuitry can be made foolproof. I would like to hear some comparison of the possibilities of protecting turn-off compo-nents in the case of short-circuit overload. From my personal experience I know that bipolar transistors can be well protected against full short circuit on the load even if they are fed from hard voltage sources. One can expect that MOSFETs have the same or even better capabilities.

F. Schwarz

I want to comment on the comparisons between transistors and thyristors because the type of inverter that I presented today has also been built with transistors. As a matter of fact NASA now builds a unit of 25 kW with transistors which have been developed especially for that purpose by Westinghouse. When I compare the transistor and the thyristor in the same application, in the same circuit, I see one significant difference and that's the ratio of the upper limit of current versus the rated current for the two devices. In a 100-A transistor you should not put through 101 A because it's rated for 100 A, whilst a thyristor rated for 100 A may be carrying as much as a 1000 A or more in one mishap. We have seen in our comparative developments that the probability of losing a switching device is much higher in transistors than in thyristors. If I wanted to rate a transistor to be comparable to the thyristor then I would have to use a 100-A transistor for 10 A. Under these conditions the transistor becomes much too small and economically almost unacceptable.

H. Stemmler (BBC Brown, Boveri & Co. Ltd., Baden)

I would like to try to answer Mr. Sittig's question. I think that while reverse-conducting thyristors will be a good solution in the short term it doesn't remove the whole commutation circuitry, so in the long term I think the gate turn-off or some other kind of turn-off thyristors will come: first for low power, then medium power, and perhaps eventually at high power ratings.

F. Schwarz

Twenty years ago I participated in a symposium at the General Electric Company. My colleagues there had devised a GTO for 200 V and 10 A at that time. Dr. Storm who was a senior consultant at the time remarked that while the performance was excellent he doubted the usefulness because he questioned how the energies in the inductances that are in series with the current would be taken care of. He pointed out that the complexity of the mechanism to turn off the SCRs could be reduced and even the associated losses might be reduced. He commented that only one part of the problem was being addressed, namely, the mechanism and the turning off of these devices, but not the energies that are carried or the magnetic fields that encircle the currents. If the currents were successfully turned off in one nanosecond, tremendous damping circuits would be needed in order to prevent destruction of the devices and the energies would have to be disposed of in one way or another. Although I believe that from the point of view of basic systems, development of the GTO is a great contribution, we should not forget that we still have to dispose of all the inductive energies that are in the circuit.

P. de Bruyne

In my opinion GTOs will not cover high-frequency applications. For a PWM inverter with 21 pulse modulation, for example, you need frequencies above 1 kHz which are difficult to realize with GTOs. We should also consider costs and these may be expressed in terms of the total silicon area of a device. For fast switching thyristors or high-voltage thyristors or reverse-conducting thyristors, the mean current density is between 50 and 70 A/cm^2. How many cm^2 do we need for a GTO to handle the same current? The result of such a discussion will be that for very low power the trigger circuit represents an important part, therefore power MOSFET's are advantageous. For medium power the commutating circuit becomes more expensive than the main thyristor and here GTOs offer an economical benefit. But for high power the commutating circuit is no more expensive relative to the total costs of an installation. The use of GTOs will not lead to lower expenses in this field.

H. Becke

According for my investigations a GTO can carry an average current density that is per total anode area of about 60-200 A/cm^2. This is somewhat less than that of a regular thyristor and therefore one can assume that the costs for a GTO will be only slightly higher than those for normal thyristors.

H. Gilgen (BBC Brown, Boveri & Co. Ltd., Baden)

Is there any experience with series and parallel connecting of GTOs?

M. Kurata

Our application engineers have successfully connected several GTOs in parallel. With respect to series connection it seems that some problems exist, but I believe these will be solved in the near future.

P. Zimmermann (Robert Bosch GmbH, Erbach)

Perhaps Mr. Kurata can come to the point he mentioned and say something about the price situation of the GTO.

M. Kurata

I wouldn't like to give a definite price but it's in the range of 1.5 times that of a usual thyristor of the same power. That is to say GTOs are not so expensive. Production yield represents a bigger problem than material costs. Therefore the surface occupied by the device is today not as important as increasing the yield in GTO production.

H.-Ch. Skudelny

So we come to the conclusion that it is to be hoped that we shall be able to buy big GTOs at almost the expense of thyristors.

K. Platzöder (Siemens AG, Munich)

I'm in charge of production and development of Siemens power devices and can give you a figure. The best large thyristors at low cost have a factor of about 1.2, which means that from 120 wafers we get 100 good devices and therefore the material costs are definitely important today. In my opinion it is not true that material costs are completely unimportant now because it's not just the silicon we have to buy it's also the capacity of the machines and things like that. The real factors today for standard thyristors are fairly good and I would like to support Mr. de Bruyne that the area you really can use also represents an important consideration.

ANALYSIS AND DESIGN OF HIGH-POWER RECTIFIERS[*]

M.S. ADLER and V.A.K. TEMPLE
General Electric Company, Schenectady, NY, USA

SUMMARY

In this paper all aspects of the design and analysis of high-power rectifiers are presented. In the first section, the results of a study are given which identify the limiting physical mechanisms affecting forward drop in power rectifiers. In the second section, the effect of packaging variations on surge and steady-state device ratings is investigated. In the third section, the effect that gold, platinum, and electron irradiation have on the switching speed, forward drop and "snappiness" of power rectifiers is analyzed using a method to directly measure the free-carrier concentration while the device is switching. Device analysis, wherever presented, is made using an exact numerical model in one dimension which allows for temperature- and time-dependent calculations on devices imbedded in inductive switching circuits.

1. INTRODUCTION

The rectifier, while ostensibly the simplest, is one of the most important and in some ways the least understood power device. For example, designing rectifiers which have the desired switching speed and the lowest possible forward drop, but not having "snappy" characteristics, in inductive switching is more of an art than a science. One of the chief problems is that while considerable effort has been put into analyzing transistors and thyristors, there has been relatively little detailed analysis of rectifiers.

In an attempt to improve this understanding, and hopefully our abilities to design rectifiers with the desired performance, the properties of power rectifiers are analyzed in this paper in a systematic fashion. The paper is divided into three major sections. In the first, the steady-state forward drop of rectifiers is analyzed in an attempt to identify the limiting physical mechanisms. This is done using an exact numerical model

---

[*] Presented at the symposium by M.S. Adler

which simultaneously solves the entire set of device equations together with the heat-flow equation in one dimension. In the second portion of the paper, the surge and steady-state properties of high-voltage rectifiers are analyzed as they are affected by packaging considerations. It is shown that substantial improvements in both steady-state and surge ratings are possible by making improvements in the device package.

This analysis is also made using the exact numerical model noted above extended to include time-dependent phenomena and the inclusion of the device package into the thermal model. In the third section of the paper we present an experimental and theoretical study of the switching performance of high-speed rectifiers where the effect on the switching speed and snappiness of the rectifier of various lifetime-control methods is investigated. The experimental technique involves the direct measurement of the free-carrier concentration profile within the device while it is switching. These results are correlated with the measured terminal properties in devices that have been either electron radiated, gold doped, or platinum doped. The same numerical model noted above is used to analyze the results. Overall the attempt of this paper is to improve the understanding of high-speed power rectifiers and to show how improved switching performance can be obtained using optimized lifetime profiles.

2. STEADY-STATE DEVICE ANALYSIS

Most details of the model used in this paper have been described in a previous publication.[1] To summarize briefly, the model is an exact numerical solution of the entire set of semiconductor-device and heat-flow equations in one dimension. A finite difference formalism based on Newton's method is used. The analysis can either be steady state or time dependent and incorporates in an integral fashion effects of heat flow and heat storage in the device and package. For modeling purpose when carrying out an analysis involving a package, the package is broken up into finite elements in the same manner as the silicon device. Among the physical effects in the model are Auger recombination,[2] Schockley-Read-Hall recombination, band-gap narrowing,[3] carrier-impurity and carrier-carrier scattering.[4]

The measured and theoretical current-voltage relations at both 300 K and 400 K are shown in Fig. 1 for the power rectifier that forms the basis for the study of this paper. The rectifier is approximately 0.012 inches thick and has a breakdown voltage of 2000 V. Fig. 2 shows the same voltage relations except that the theoretical and experimental curves are plotted together to show the crossover in the characteristics. As can be seen the model is capable of predicting the I-V curves at both 300 K and 400 K including the crossover in the measured curves. The latter refers to the fact

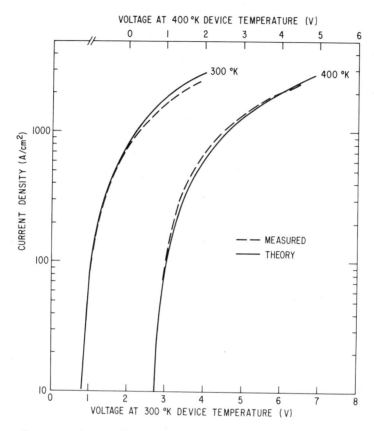

Fig. 1   Current-voltage relation for high-voltage rectifiers at 300 K and 400 K.

that the forward drop is lower at 400 K than 300 K for currents less than approxima-tely 300 A cm^{-2} and greater for higher currents.

In order to achieve this level of accuracy, it was necessary to carefully model the temperature dependence of all the device parameters such as the band gap and the carrier mobilities. The temperature dependence for these parameters was taken from a paper by Slotboom and DeGraaff.[3] The significance of these results is that it becomes possible to predict accurately device behavior under all forms of excitation including surge and steady-state operation. With this corroboration of the accuracy of the model, we now apply the analysis to the investigation of the factors which limit steady-state device performance.

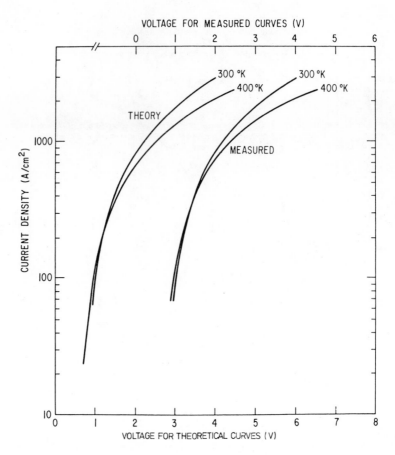

Fig. 2   Current-voltage relation for high-voltage rectifier plotted to show the cross-over in the characteristics.

Fig. 3 is a series of calculated I-V curves for a device similar to that shown in Fig. 1 and with a heat-sink temperature of 400 K. Such a high temperature was chosen for this analysis since the device is often at elevated temperatures during operation and an understanding of limiting physical effects is most appropriately done at this temperature. The "temperature is constant" curve is the result of holding the device at a constant temperature of 400 K. With a heat-sink thermal conductance of G = 1.35 W cm^{-2}/K, which is an average value for a press pack, the I-V curve changes dramatically with a device temperature rise of 90 K at the 2-V level.

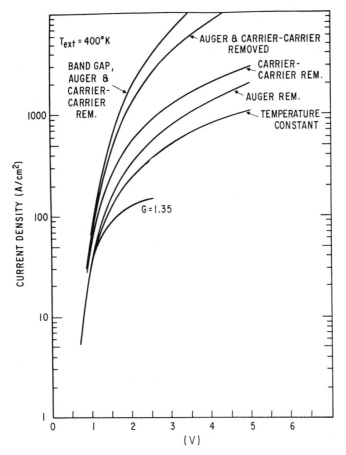

Fig. 3   Theoretical curve illustrating the relative importance of various physical mechanisms, including heat flow, on the I-V characteristics.

The remaining curves are associated with an investigation to determine the physical
phenomena which limit device performance. The temperature is held constant at 400 K
for this study. Each curve is labelled to indicate the physical mechanisms that were
removed: "carrier-carrier" for example, indicates the removal of carrier-carrier scattering. The 1000-A cm^{-2} level is associated with maximum surge operation while the
100-A cm^{-2} level is associated with maximum steady-state operation. As can be seen,
carrier-carrier scattering and Auger recombination are important limiting mechanisms
at the surge operating levels and also at the 100-A cm^{-2} level. There was virtually no
effect of removing band-gap narrowing at any level until both Auger recombination and
carrier-carrier scattering had been removed.

The mechanism of SRH recombination, not mentioned so far, becomes the limiting factor at levels below 100 A cm^{-2}, and is also of great importance at the surge levels. This can be seen in Fig. 4 which shows the integrated recombination rate for two levels of operation corresponding to surge and steady-state operation. The surge level curves (dashed) are for a 5.0-V forward drop while the steady-state curves (solid) are for a 1.25-V forward drop. Also note the different set of axes that are used for each level. Two curves are shown for each level in order to distinguish the various recombination mechanisms at work in the device. The curve labeled "total recombination" includes Auger, SRH, and surface recombination, while that labeled "SRH only" includes only the SRH mechanism. It is thus seen that at the 1.25-V level the SRH mechanism accounts for most of the bulk recombination, amounting to 90 % of the total. At the 5.0-V level, Auger recombination has become of substantial importance, as was seen in Fig. 3, but still only accounts for 30 % of the total bulk recombination. It is also immediately apparent that the vast majority of the recombination at both levels of

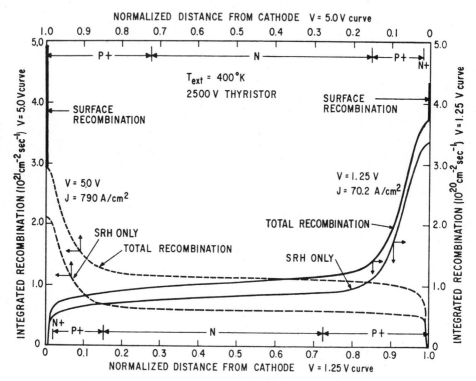

Fig. 4   Integrated recombination rate under maximum steady-state and surge conditions.

operation occurs in the ends of the device with the anode end having by far the largest contribution. At the 5.0-V level 77 % of the total recombination occurs within the p^+ anode region and 21 % within the p^+-n^+ cathode regions. These overall conclusions are consistent with the findings of other workers.[5]

The surface recombination at the anode end of the device is due to minority electrons recombining at the anode contact. As can be seen, the size of the effect, particularly at the 5.0-V level, is quite substantial. At this level, almost half of the current reaching the anode is electron current. It should be noted, however, that the amount of surface recombination shown is a function of the assumed dependence of lifetime on doping concentration discussed earlier. Should the electron lifetime be lower than assumed in the p^+ anode regions, fewer electrons would reach the anode contact thereby reducing the size of the surface recombination. This large size of electron surface recombination, while possibly being exaggerated, serves to illustrate the role of surface recombination in contributing to power dissipation, as is seen below. The situation for minority holes at the cathode is quite different with virtually no hole current reaching the cathode even at the highest levels. This is a result of the lower lifetime and mobility for holes as compared to electrons together with the fact that the cathode is doped to a much higher level than the anode.

2.1 Heat Dissipation

Fig. 5 shows the integrated power dissipation density assuming the device was at a constant temperature T = 300 K. The two curves shown correspond to steady-state (solid) and surge conditions (dashed). Also note the different set of axes that are used. In both curves one can see the effect of recombination at the contacts, particularly the effect of minority electrons recombining at the p^+ anode contact (X = 1). The main distinction between the two cases is the fact that for the 1.25-V level, 75 % of the heat is dissipated outside the n-base region while at the 5.0-V level, over half of the heat is dissipated in the n-base region. Further insight into this can be obtained by comparing Fig. 4, which shows the integrated recombination, with Fig. 5, which shows the integrated power density. Although Fig. 4 was for a heat-sink temperature of 400 K, it is qualitatively similar to the 300-K curve. As can be seen, the 1.25-V curve in Fig. 5 is somewhat similar to the 1.25-V curve in Fig. 4, indicating that recombination plays a large role in producing the power dissipation at this level. One can see the significant role that surface recombination has in producing power dissipation. However, the high level curve in Fig. 5 bears little resemblance to the integrated recombination curve in Fig. 4. Comparison of the curves indicated that at surge levels the power dissipation is much more uniformly distributed throughout the device than the recombination which mainly occurs in the ends of the device. Thus it can be concluded that ohmic heating produces most of the power dissipation in the bulk of the device although surface recombination can still be seen to contribute to the power dissipation in the anode region.

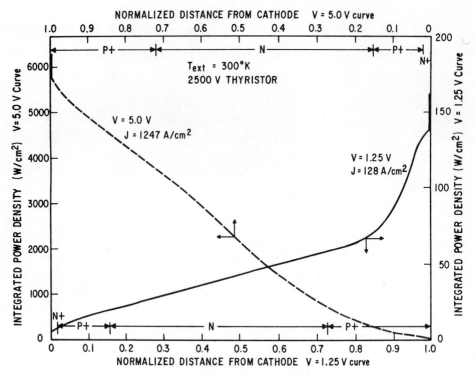

Fig. 5   Integrated power dissipation under maximum steady-state and surge conditions.

2.2 Summary

In this section a study has been undertaken to identify the limiting physical mechanisms operating in power-rectifier devices. It was shown that SRH recombination is an important mechanism over the range of device operating conditions for maximum steady-state to surge levels. Auger recombination and carrier-carrier scattering effects become important at the surge levels. Band-gap narrowing was not found to have much effect for any of the devices under study. It was also shown that recombination occurs mainly in the ends of the device with a large amount of surface recombination occurring at the p^+ anode at surge levels. However, it was shown that device temperature rise due to thermal resistance on the package is potentially the largest factor limiting device performance. This is discussed further in the next section.

An attempt was made to correlate the integrated power dissipation density in the device with the integrated recombination rate. It was shown that at maximum steady-state levels power dissipation occurs mainly as a result of carrier recombination in the bulk of the device while at surge levels power dissipation becomes dominated by the effects of ohmic heating. However, surface recombination at the anode was shown to add significantly to the total device power dissipation at surge levels. This, together with the high rate of recombination that occurred in the anode, resulted in the fact that 66 % of the power was dissipated in the anode half of the study device.

3. STANDARD PRESS PACK ANALYSIS INCLUDING DEVICE PACKAGE

Having investigated the physical mechanism limiting device performance, in this section of the paper the effect that the device package has on the overall device ratings is analyzed.

Fig. 6 gives details of the device press package that forms the basis for this section of the paper. The package dimensions are shown on one side of this figure and the

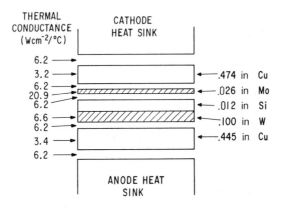

Fig. 6   Schematic drawing of the regular package.

thermal conductances for the elements in the package on the other. The values for these conductances were calculated using the data shown in Table 1. Also shown in the table are the values for the bulk thermal capacities. Referring to the package elements, the 0.026-inch molybdenum plate is the strain buffer, the 0.010-inch tungsten plate is the backup plate for the silicon device and is alloyed to the silicon, and the thick copper pieces represent the bulk of the package. For each of the five dry interfaces in the package a value of thermal conductance of 6.2 $W \cdot cm^{-2} K^{-1}$ is used. This figure was arrived at using the thermal resistance data in the General Electric Semi-

TABLE 1

Thermal Properties

Material	Thermal Conductivity $(W \cdot cm^{-1} K^{-1})$	Thermal Capacity $(W \cdot s \cdot cm^{-3} K^{-1})$
Copper	3.82	3.42
Molybdenum	1.38	2.70
Tungsten	1.67	2.67
Lead	0.346	1.47
Silicon	$3110/T^{4/3}$	1.68 T = 300 K
		1.8 T = 350 K
		1.87 T = 400 K
		1.93 T = 450 K

conductor Data Book for the C701 device (not the device here but a similar package). The value of the bulk thermal resistance for each of the package elements is subtracted from the overall d.c.-junction to heat-sink thermal resistance and the remainder divided among the five dry interfaces. The dry interfaces are treated as thermal resistors with no thermal mass while each of the other package elements have both. The thermal resistance for the alloy interface between the tungsten and the silicon is assumed to be negligible.[6,7]

The measured and predicted temperature rise at the end of a half-wave 60-Hz sinusoidal current surge are shown in Fig. 7 as a function of the peak surge current density for the device discussed in the context of Fig. 1 and for the standard press pack shown above. The temperature is measured 1 ms after the half-cycle current surge by passing a test current through the device and measuring the voltage. The temperature is then determined from a calibration curve measured previously on the same device. As such, the temperature measured is the junction temperature. However, because of the fact that at 60 Hz the silicon device itself is in thermal equilibrium and the fact that the bulk of the thermal resistance and mass is external to the device, the temperature across the silicon device is uniform to within 15 % (see Fig. 8). In any case, the theoretical curve shown in Fig. 7 and all theoretical data elsewhere in the paper are for the peak device temperature, which is within several degrees of the junction temperature. As can be seen, the theoretical and experimental curves are in good agreement, again providing strong evidence of the accuracy of the electrical and thermal modeling for both the silicon device as well as the package. Using this

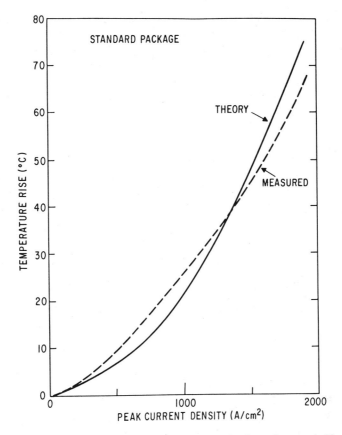

Fig. 7   Temperature rise for the regular package device after a half-wave 60-Hz current pulse as a function of the peak current density.

as a basis, a theoretical study of the effect that packaging variations have on both surge and steady-state performance is now presented.

3.1 Surge Analysis

Fig. 8 shows the temperature rise in the silicon and in the package as a function of position for several different times during the half-wave 60-Hz surge. It should be noted that the distance scale is expanded in the silicon. The effects of the dry interfaces on the cathode side of the device (zero of the graph) are immediately apparent in the form of abrupt temperature drops. It can also be seen that the significant temperature excursions in the tungsten backup plate are confined to within 0.030 inches of the silicon. The properties of the rest of the package do not affect the surge oper-

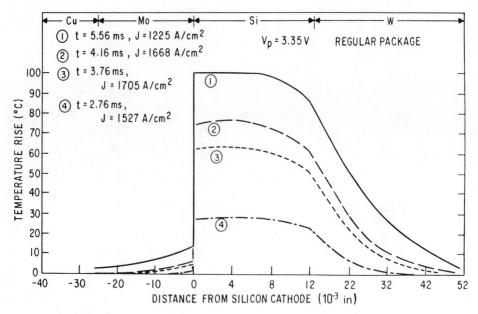

Fig. 8   Temperature profile in the regular package at several times in the half-wave
          pulse.

ation. Another interesting observation is that the peak current flow occurs at t =
3.76 ms which is before the peak in the applied voltage (t = 4.16 ms). This offset
is due to the increased forward resistance of the device as it heats up during the
surge. The peak in the device temperature occurs at t = 5.56 ms which is considerably after the voltage peak. This delay occurs because of the thermal time constant
of the package materials nearest the silicon.

Fig. 9 shows the temperature rise after the half-wave sinusoidal current surge for
four devices which incorporate a series of variations on the regular package. Curve 1
is identical with the theoretical curve in Fig. 7 for the regular package device. Curve
2 is for a device where the strain buffer has been removed. As can be seen, there is
little difference since the improvement in the thermal properties of the copper (see
Table 1) over the 0.026-inch molybdenum strain buffer is almost completely masked by
the single dry interface that is still present. Curve 3 is for a device where the dry
interface between the strain buffer and the silicon is removed and, as is seen, there
is a considerable reduction in the temperature rise. Curve 4 is for a device where the
strain buffer is removed and where there is a perfect interface with the copper package. Curve 5 is for a device similar to that for curve 4 but where the anode tungsten
backup plate is replaced by one of copper.

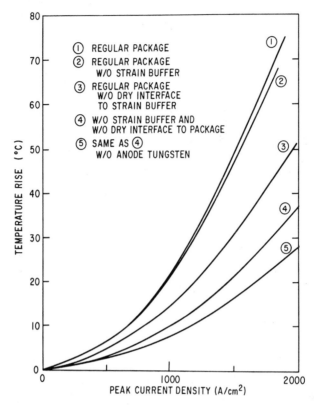

Fig. 9   Temperature rise for several package types after a half-wave 60-Hz current pulse as a function of the peak current density.

Fig. 10 shows a comparison of the temperature profiles for three of the package types discussed above under surge excitations that produce peak temperature rises of 100 K. The first curve is a duplicate of the t = 5.56 ms curve shown in Fig. 8 for the regular package. The second curve is for a device where the dry interface between the silicon and the strain buffer is removed. As a result, the device is capable of handling a 26 % increase in surge current and a 40 % increase in power over that of the regular package device before reaching the 100 K temperature rise. The third curve is for a device with no strain buffer and a perfect interface between the package and the silicon. This device shows a 38 % increase in current capability and a 59 % increase in power-handling capability over the regular package device. As can be seen from curves 2 and 3, the effect of having a perfect interface is to eliminate abrupt thermal drops and to lower the temperature on the cathode side of the silicon wafer. The rea-

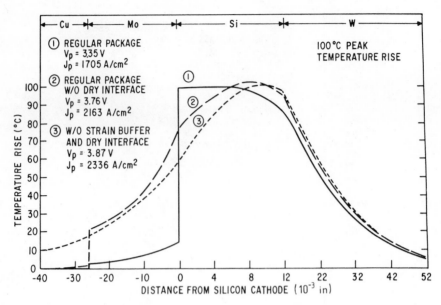

Fig. 10   Temperature profile in the three package types 5.56 ms into a 60-Hz current
surge under conditions that produce a 100 K peak temperature rise. Note
the difference in distance scale within the silicon wafer.

son that the tungsten backup plate is less effective in lowering the temperature on
the anode side of the wafer is that most of the power dissipation ($\sim$ 70 %) is on the
anode half of the device (see Fig. 5).

It is also worth noting that the temperature rise is largely confined to the 0.030 inches
of the package adjacent to the silicon wafer during the current surge. The heat simply
has not had the time to propagate beyond this point. Thus one can see that with respect to surge behavior only the properties of the package nearest the silicon wafer
are of importance and the single dry interface at the silicon cathode has an exception-
ally large effect. A comparison of the device types discussed above, showing the peak
current and voltage to reach a device temperature of 100 K, is summarized in Table 2.
The columns indicated as "% J" and "% P" represent the percentage increase in the
surge current and power for the indicated device over the regular package. The first
five device types are the same as discussed in Fig. 9, while the last is for a device
where the surfaces of the silicon wafer are cooled to hold their temperature at 300 K.
This would obviously represent the ultimate in packaging technology.

As can be seen, a substantial improvement in device performance can be achieved by
improving the thermal properties at the interface between the silicon device and the
first package layer. Even a soldered interface goes a long way in doing this. As noted

earlier, the 0.026-inch molybdenum strain buffer is not a significant problem in itself and only becomes an issue once the first dry interface is removed. Of course, if the strain buffer were substantially thinner then it would become a problem since the dry interface between the strain buffer and the copper end pieces would begin to inhibit the heat flow. For the 0.026-inch buffer in this study, this dry interface does not have a significant effect during the 60-Hz surge (see Fig. 8).

TABLE 2

Half-Wave Sinusoidal Surge

(Peak temperature rise of 100 K is the criterion)

	V_p (V)	J_p (A cm^{-2})	%J	%P
1. Regular package	3.34	1698		
2. Regular package without strain buffer	3.35	1700		
3. Regular package with 0.003-inch lead instead of dry interfaces	3.65	2001	18	29
4. Regular package without dry interface to strain buffer	3.72	2133	26	40
5. Without strain buffer and without dry interface to package	3.87	2336	38	59
6. Contacts cooled	6.5	5270	310	604

Several devices have been fabricated which approximate the improvements indicated in Table 2. Fig. 11 shows the temperature rise after a half-wave 60-Hz surge as a function of peak current density for three devices. In each case both a theoretical and an experimental curve is shown. The upper pair of curves are identical to those of Fig. 7 for the regular device with the molybdenum strain buffer. The middle pair of curves is for a device where the silicon device is soldered to the cathode end piece. The theoretical curve here assumes a 0.003-inch lead interface. As can be seen, the measured curve lies above the theoretical one indicating that either the solder joint is thicker than 0.003 inches or that it has not made good contact with the silicon. There is some evidence supporting the latter interpretation. The third set of curves is for a device with a thin metallurgical interface. The theoretical curve is actually for a device with a perfect interface. As can be seen, the experimental device performs almost as well as the "perfect" theoretical device.

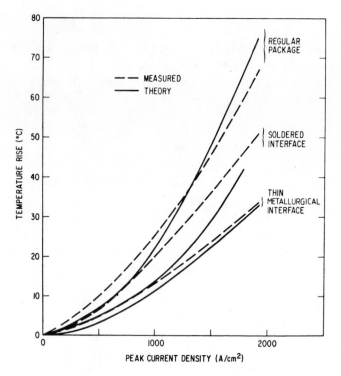

Fig. 11   Temperature rise for devices with a regular package, a soldered interface to
the silicon, and a thin metallurgical interface to the silicon after a half-wave
60-Hz current pulse as a function of the peak current density.

3.2 d.c. Analysis

Fig. 12 shows temperature profiles for the three device types shown in Fig. 10. However, in the present case, the profiles are for d.c. steady-state excitations that produce the 100 K peak temperature rises. As was the case in Fig. 10, there is a difference in scale within the silicon wafer. In the first curve the two dry interfaces on
each side of the strain buffer are clearly seen, as is the dry interface between the
copper package and the tungsten backup plate on the anode side. The differences in
the slopes of the temperature profiles are indicative of the differences in the thermal
conductances of each of the package materials (molybdenum, tungsten, copper). The
second curve is for the case where the dry interface between the silicon and the
strain buffer has been removed. Removing this single dry interface has resulted in
an 8 % increase in d.c. current and an 11 % increase in power dissipation over the
regular package for the same 100 K temperature rise. This can be contrasted to the
26 % and 40 % increases in surge current and power-handling capability observed by
removing the same dry interface. (See Table 2 and Fig. 10.) The reason for the difference is easily explained. The interface in question is critically important on the
surge properties but is only one of five interfaces that affect steady-state perfor-

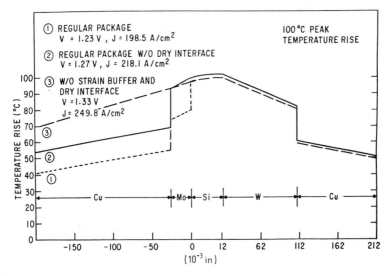

Fig. 12   Temperature profile in three package types under d.c. steady-state conditions that produce a 100 K peak temperature rise. Note the difference in
distance scale within the silicon wafer.

mance. The third curve is for the case where the strain buffer is removed and where
there is a perfect interface between the silicon wafer and the copper package. Again
the improvement is less than the case for the same change in the surge analysis.

Table 3 shows an overall summary for all of the device types previously discussed in
the surge analysis as will as three others. The first four are the same devices shown
in Fig. 9 and Cases 1, 2, 4 and 5 in Table 2. In all cases, the percentage in the
power-handling capabilities is considerably less than shown in Table 2. The last case
is for the same device as the last entry in Table 2 where the surfaces of the silicon
wafer are held at 300 K. In this case, the current and voltage are the same as for
the surge case since the silicon is able to stay in thermal equilibrium at the 60-Hz
rate.

It is interesting to note that the improvement for this perfectly cooled device over the
regular package seen here is an order of magnitude greater than the corresponding
improvement in the surge properties. The basic reason for this, as explained above,
is that the surge properties are determined mainly by the silicon device and the first
0.030 inches of the package adjacent to the silicon. In the steady-state case, every
inch of the package affects performance, greatly adding to the overall thermal resistance of the device. It can thus be seen that the largest overall improvement in device
performance is to be found in the steady-state characteristics, although some specific
changes have a greater effect on the surge ratings.

TABLE 3

d.c. Steady State

(Peak temperature rise of 100 K is the criterion)

	V (V)	J (A cm^{-2})	%J	%P
1. Regular package	1.232	199	-	-
2. Regular package without dry interface to strain buffer	1.263	215	8	11
3. Regular package without dry interface to strain buffer	1.276	222	12	16
4. Without strain buffer and without dry interface to package	1.33	250	26	36
5. Contacts cooled	6.5	5270	2458	13872

3.3 Summary

Dramatic improvements in both the surge-current ratings and the steady-state-current ratings have been shown to be achievable through packaging and cooling innovations. Specifically, surge-current ratings could potentially be increased by a factor of three over the regular package device and the steady-state-current ratings by a factor of 25. Achieving these dramatic improvements will not be easy since it amounts to cooling the silicon surface to the ambient. However, significant improvements, on the order of 50 % in surge ratings and 90 % in steady-state ratings, could be achieved by eliminating the nonmetallurgical (dry) interfaces in the package.

It is recognized, of course, that the thermal and electrical properties are not the only important characteristics affecting device performance. Mechanical problems, such as the scrubbing of the silicon device against the package and work hardening of soldered interfaces, are real limits to device longevity and reliability. Clearly, the challenge is to maintain the mechanical integrity while striving for improved thermal and electrical characteristics. However, it is also true that all three sets of properties are related. Reductions in power dissipation leading to smaller thermal cycles will also minimize the adverse mechanical effects. It has been the purpose of this section to show the size of the potential improvement in the electrical and thermal characteristics which could be expected for a wide range of packaging variations.

4. USE OF GOLD, PLATINUM, AND ELECTRON RADIATION
TO CONTROL SWITCHING SPEED

In this section of the paper, the results are presented of a study to investigate the effect on the reverse recovery time t_{rr} and the so-called "snappiness" of the turn-off of various techniques for controlling carrier lifetime. The latter is defined as the ratio $S = t_B/t_A$ shown in Fig. 13, a device with a small ratio being snappy in that it produces undesirable voltage spikes in inductive switching. Platinum and gold diffusions as well as electron irradiation are investigated using a free-carrier infrared absorption technique to measure quantitatively carrier profiles in both the steady state and during switching transients. This technique is applied to produce measurements of the carrier distribution in power rectifiers as a function of injection level and during open-circuit transients. The exact numerical computer model described earlier is then used to analyze the data. A more complete presentation of the results and a detailed description of the technique is given in an earlier publication.[8]

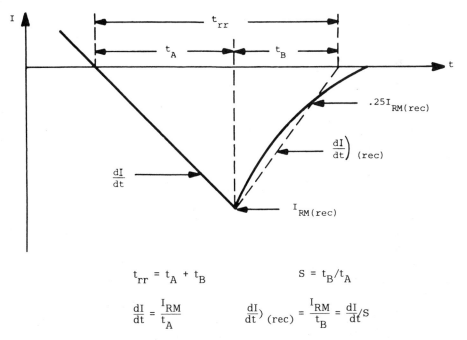

$$t_{rr} = t_A + t_B \qquad\qquad S = t_B/t_A$$

$$\frac{dI}{dt} = \frac{I_{RM}}{t_A} \qquad\qquad \frac{dI}{dt}\bigg) (rec) = \frac{I_{RM}}{t_B} = \frac{dI}{dt}/S$$

Fig. 13 . Schematic representation of a reverse-recovery waveform and definition of terms.

4.1 High-Lifetime Sample

The device under investigation is a p^+-p-n^--n^+ rectifier 350-μm thick, with 1 x 10^{14} cm^{-3} substrate doping (n region), and an 8- μm p^+ region with a 10^{20} cm^{-3} surface concentration, an 80-μm p region with a 2 x 10^{17} cm^{-3} surface concentration, and a 20-μm n^+ region with a 10^{20} cm^{-3} surface concentration. Fig. 14 shows the measured steady-state carrier distribution for three different injection levels corresponding to anode currents of 870, 435, and 109 A cm^{-2}. Theoretical curves are also shown and these are discussed below.

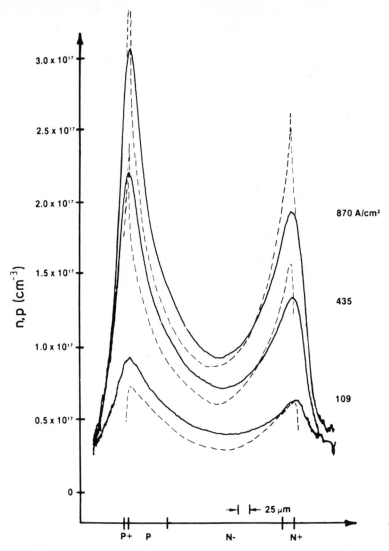

Fig. 14   Measured steady-state distributions (solid lines) and computer-simulated curves (dashed) at 870, 435, and 109 A cm^{-2}, for the high-lifetime sample.

The method also makes it possible to measure the carrier distribution during device switching. This is illustrated in Fig. 15 for the same sample as that in Fig. 14 (referred to subsequently as the high-lifetime sample). During switching, initially both diffusion and recombination take place, and only after the carrier distribution has flattened out does it decay as a whole. Once this condition is reached, these curves can be used directly to determine the carrier lifetime in the n⁻ region. In the present case, a 9-µs high-level lifetime is measured. The fact that the carrier level appears to fall below zero in the heavy-doped regions is due to electronically taking the logarithm of a ratio of two signals that are very small and which differ by only a very small amount.

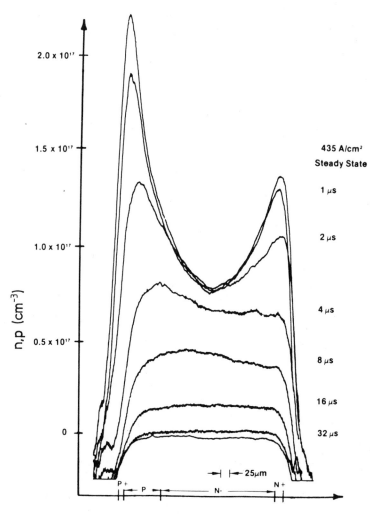

Fig. 15   Measured open-circuit decay curves for the high-lifetime sample.

As noted earlier, the measured steady-state free-carrier distribution is shown in Fig. 14 at three different injection levels for the high-lifetime sample. This figure also shows the calculated set of free-carrier distributions at these injection levels using the same exact one-dimensional model described earlier. The major unknown in the analysis of this experiment and, in fact, in most semiconductor device analyses, is the value of the local carrier lifetime. The doping profile can be determined from spreading-resistance measurements and the carrier moblities have been well characterized in the literature. However, the local lifetime is a function of processing and starting material, and is also difficult to measure. It was one of the goals of this experiment to determine the lifetime profile.

Although the experiment provides a great deal of information, it does not directly provide the ratio the low-level lifetimes, nor does it directly yield the recombination lifetimes in the heavily doped regions of the device. These parameters, however, strongly affect the resulting carrier distribution. The low-level electron lifetime in the p^+n region and the low-level hole lifetime in the n^+ region determine the injection efficiencies of the p^+-p-n^- and the n^+-n^--p junctions, respectively. These injection efficiencies, in turn, directly affect height of the p^+ and n^+ peaks of the carrier distribution as well as the modulation level in the n^- region.

Unfortunately, the heights of the p^+ and n^+ peaks and the n-modulation level are not sufficient in themselves to determine the dependence of carrier lifetime on doping. The best that is possible is to determine an average lifetime for each of the heavily doped end regions. However, additional information can be obtained from the distribution of the free carriers in the heavily doped regions themselves. While the small size of the p^+ and n^+ regions relative to the size of the beam (25 μm) makes it difficult to use the data quantitatively in these regions, the experiment does show a large number of free carriers in the p region adjacent to the p^+ anode. Furthermore, the modeling results show that the carrier distribution in this region is a sensitive function of the dependence of lifetime on doping. As a result, it was possible to determine the dependence of lifetime on doping as well as to determine the ratio of low-level electron and hole lifetimes using a fitting procedure based on the measured carrier distribution and the current-voltage relation.

In the analysis of the measured carrier distributions in Fig. 14, the sum of the low-level electron and hole lifetimes were set equal to 9 μs in the n region as determined by the open-circuit decay measurements shown in Fig. 15. The value of the low-level electron lifetime was taken as eight times that of the hole lifetime yielding a value of 8 μs for the electron lifetime in the n region. The basis for this choice was the fact that the relative size of the p^+ and n^+ peaks in the calculated steady-state free-carrier distribution is a weak function of the hole-electron lifetime ratio and the value of eight gave the best overall results. The values of the lifetime in the heavily doped region were set equal to the values in the n^- region multiplied by the three-tenths power of the ratio of the doping in the n^- region to the doping at the point in question. This choice yields a reasonable fit to the measured free-carrier distribution in the p^+-p region as well as reasonable agreement with measured values of the current-voltage relationship. Values for the exponent in the scaling law greater than three tenths caused the calculated distribution to fall off too rapidly with distance from the p^+-p junction and raised the forward drop above measured values. Values less than this resulted in slightly better agreement with the measured injected-carrier distribution, but lowered the forward drop. It should also be noted that good results have been obtained in predicting terminal characteristics for a large variety of power rectifiers and thyristors using this same inverse 0.3 power of lifetime on doping.[1,9,10]

Referring to Fig. 14, although the agreement between the calculated and measured carrier distributions is within 5 % for most of the 870-A cm^{-2} curve, the disagreement is significant for the p^+ and n^+ peaks. It is probable that the discrepancy is due to the fact that the measurement is not capable of resolving such narrow peaks since the laser beamwidth is 25 μm at the focal point. The agreement between theory and experiment is not as good for the lower curves, dropping to 14 % for the 435-A cm^{-2} curve and 25 % for the 109-A cm^{-2} curve. However, the overall agreement is quite reasonable considering the uncertainties in the measurement and the values of lifetime used in the theory.

4.2 Effects of Various Means of Controlling Lifetime

Three methods of controlling lifetime have been investigated: electron irradiation, gold diffusion, and platinum diffusion. A composite of measured steady-state distributions for samples using each of the lifetime-control methods is shown in Fig. 16. The primary observations which can be made from these curves are that: a) as expected, lifetime killing reduces the level of injected charge within the lightly doped region; b) the samples doped with gold and with platinum exhibit a peak reversal with the peak on the n^+ side becoming the dominant one; and c) the n^+ peak is larger for the platinum-doped sample than for the high-lifetime sample. The second of these observations may be influenced by the fact that the source for both the gold and the platinum diffusion was a coating on the p^+ side of the wafers with the result that the lifetime has been reduced more significantly on the p^+ side than on the n^+ side. This will be discussed in greater detail below.

The third observation concerning the n^+ peak in the platinum-doped sample warrants further investigation. The behavior of the charge distribution in the platinum-doped samples after open circuiting the device is shown in the experimental plots in Fig. 17. Here we see that the decay rate is substantially slower on the n^+ side of the sample, indicating a higher lifetime in that region. This could explain the elevated n^+ peak as well as the long turn-off tails for platinum-doped samples which was observed previously.[11]

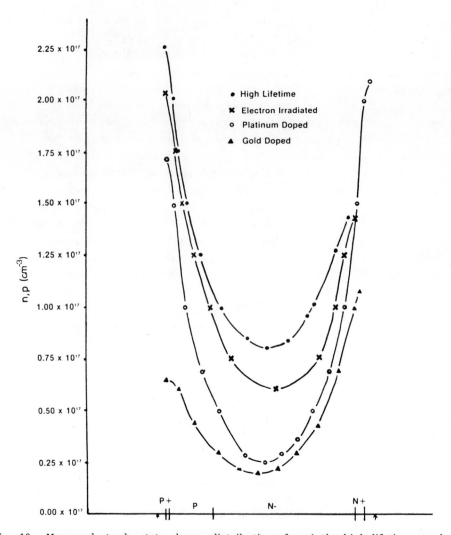

Fig. 16   Measured steady-state charge distributions for a) the high-lifetime sample at
          435 A cm^{-2}; b) the same sample after electron irradiation and annealing,
          also at 435 A cm^{-2}; c) the platinum-doped sample at 460 A cm^{-2}; and d) the
          gold-doped sample at 400 A cm^{-2}.

The reverse-recovery measurements of these samples are summarized in Table 4. Here
t_{rr} is broken down into its components t_A and t_B which yields a different "snappiness" ratio for each device. Here t_A is the time required for the blocking p-n junction
to come out of saturation while t_B is the time required to "sweep out" the remaining
charge. The shorter t_B is relative to t_A the "snappier" the device.

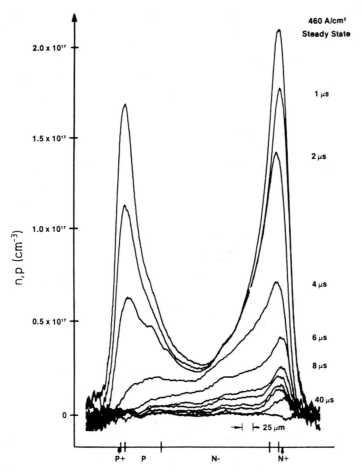

Fig. 17   Measured open-circuit decay curves for the platinum-doped sample.

TABLE 4

Treatment	t_A	t_B	t_{rr} $(t_A + t_B)$	S $(t_A + t_B)$	dI/dt	V_f	I_r
	(μs)	(μs)	(μs)	-	(A μs^{-1})	(J_f = 85 A cm^{-2})	(μA)
None	18.5	2.8	21.3	.16	24	1.02	70
Au doped	2.3	1.3	3.6	.56	24	1.25	50
Pt doped	1.6	1.0	2.6	.62	24	1.46	-
e$^-$ irradiated 5x10^{13} cm^{-2}	3.5	1.2	4.7	.34	24	1.2	40

As can be seen, the platinum-doped sample is the least snappy, with the gold doped, electron irradiated, and high-lifetime samples following in order of increasing "snappiness". Little about the effect of the lifetime treatment on turn-off time itself can be concluded from Table 4 since the relative levels of lifetime changes are different for each case. However, Fig. 18 shows the tradeoff between turn-off time and forward drop for all of the cases in Table 4. While even this is difficult to interpret, to first order it appears that all four cases fall on the same curve indicating that there is little difference in the methods with respect to simply shortening the turn-off time. The data for leakage current are incomplete but it appears that the gold-doped sample produces the highest leakage current even though the platinum sample had a lower turn-off time.

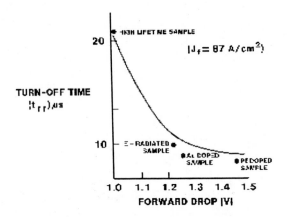

Fig. 18   Graph of switching speed vs. forward drop for high-lifetime sample, gold- and platinum-doped samples and E-beam irradiated sample.

The model was then applied to determine the effect of the lifetime-control methods on the carrier-lifetime profiles. In the case of the electron-irradiated sample good results were obtained by leaving the carrier-lifetime profiles the same in the heavily doped regions as for the high-lifetime sample but reducing the lifetime in the n region from 9 μs to 6.3 μs, the latter being determined from the open-circuit decay measurements. This resulted in good agreement between theory and experiment for the carrier-profile measurements as well as in measurements of the I-V relation. This combined agreement provides substantial confirmation of the effect of electron irradiation on the carrier-lifetime distribution noted above.

No attempt was made to do this type of complete analysis for the gold- and platinum-doped samples since the results in Fig. 16 show that a complicated alteration of the carrier lifetimes has taken place. This is particularly true in the case of the platinum-doped sample. The p^+ peak is seen to be approximately unchanged, but the modulation level has been greatly lowered in the n^- region (by a factor of three). The situation for the gold-doped sample is not quite as complicated: here it is seen that the p^+ peak has been greatly reduced together with a lowering of the modulation level in the n^- region. This can be approximately accounted for analytically by lowering the lifetime in the p^+-p region by a factor of four.

4.3 Summary

In this section, a free-carrier infrared absorption technique has been applied to a p^+-n-n^--n^+ power rectifier. By analyzing both the steady-state and open-circuit transient data it was determined that the carrier lifetime varies inversely as the 0.3 power of the doping. The technique was also used to analyze devices which had been either electron irradiated or had gold or platinum diffusion done on them. Based on terminal measurements, it was determined that the platinum-doped sample had the least snappy behavior followed by the gold-doped, the electron-irradiated and the untreated samples. The carrier distributions measured by the infrared absorption technique revealed that the effect of the electron irradiation was to uniformly reduce the lifetime in the lightly doped n^- region. The lifetimes in the heavily doped regions were not appreciably affected. In the case of the gold-doped sample, the effect could be approximated by reducing the lifetime in the p^+-p region by a factor of four, this being the side of the device which was the source of the gold diffusion. The platinum device was the most complex where the lifetime in the p^+-p region on the side where the diffusion was done was reduced, but not as much as the gold sample. In the n^+ region, the lifetime seems to be increased. The major lifetime-killing effect was to reduce the lifetime in the n^- region by a factor of three. To explain this strange lifetime distribution a thorough analysis of the dependence of recombination rate on doping and free carrier concentrations has to be carried out. Since both the gold and platinum devices show a relative decrease in lifetime in the p^+ side of the device versus the n^+ side and both have soft or unsnappy turn-off behavior it is to be assumed that there is some correlation.

5. CONCLUSIONS

In this paper all aspects of the design and analysis of high-power rectifiers have been presented. This includes the identification of the limiting physical mechanisms affecting forward drop, the effect that packaging variations have on the surge and steady-state ratings, the effect that gold, platinum and electron radiation have on the switching speed and snappiness of power rectifiers, and finally a method for improving the tradeoff between forward drop and switching time. In the first section it was shown that carrier lifetime and carrier-carrier scattering are important limiting effects on the forward drop. However, it was clear that the largest limiting mechanism was device temperature rise due to device heating. This theme was amplified in the second section where it was shown that both the surge and steady-state current and power-dissipation ratings can be dramatically increased by improvements in device packaging. As an example, while surge-current ratings potentially can be increased by 300 % if the device contacts are held at ambient temperature, a more realistic improvement of 40 % can be achieved by eliminating the nonmetallurgical interfaces in the standard press pack. A 25 % improvement in the steady-state ratings can also be made by this change. In the third section it was shown that all of the lifetime-control methods (gold and platinum doping and electron irradiation) produce devices that are less snappy than an untreated device. However, the platinum-doped sample had the least snappy behavior while also having the lowest increase in leakage current of any of the methods. All of the methods had the same approximate tradeoff between switching speed and forward drop.

REFERENCES

1. M.S. Adler, "Accurate Calculations of the Forward Drop of Power Rectifiers and Thyristors," IEEE Trans. Electron Devices, ED-25, (January 1978) 16-22.

2. N.G. Nilsson and K.G. Svantesson, "The Spectrum and Decay of the Recombination Radiation from Strongly Excited Silicon," Solid State Commun., 9 (1972) 155-159.

3. J.W. Slotboom and H.D. DeGraaff, "Measurements of Bandgap Narrowing in Si Bipolar Transistors," Solid-State Electron., 19 (1976) 857-862.

4. D.M. Caughey and R.E. Thomas, "Carrier Mobilities in Silicon Empirically Related to Doping and Field," Proc. IEEE (Lett.), 55 (1967) 2192-2193.

5. J. Burtscher, F. Dannhauser, and J. Krausse, "Rekombination in Thyristoren und Gleichrichtern," Solid-State Electron., 18 (1975) 35-63.

6. F.W. Staub and M.H. McLaughlin, "High Current Solid State Power Conversion Equipment Cooling," paper AlChE-15 presented at National Heat Transfer Conference, August 1975.

7. M.H. McLaughlin and E.E. Von Zastrow, "Power Semiconductor Equipment Cooling Methods and Application Criteria," in Conf. Rec. 1974 9th Ann. Meet. IEEE Industry Applications Soc. (October 1974), 1119-1129.

8. D.E. Houston, M.S. Adler, and E.D. Wolley, "Measurement and Analysis of Carrier Distribution and Lifetime in Fast Switching Power Rectifiers," IEEE Trans. Electron Devices, ED-27 (1980) 1217-1222.

9. M.S. Adler, B.A. Beatty, S. Krishna, V.A.K. Temple, and M.L. Torreno, "Second Breakdown in Power Transistors Due to Avalanche Injection," IEEE Trans. Electron Devices, ED-23 (1976) 851-857.

10. M.S. Adler and H. Glascock, "Surge Characteristics of Power Rectifiers and Thyristors," IEEE Trans. Electron Devices, ED-26 (1979) 1085-1091.

11. M.D. Miller, "Differences between Platinum- and Gold-Doped Silicon Power Devices," IEEE Trans. Electron Devices, ED-23 (1976) 1279-1283.

12. J.M. Fairfield and B.V. Gokhale, "Gold as a Recombination Center in Silicon," Solid State Electron., 8 (1965) 685.

13. K.S. Tarneja and J.E. Johnson, "Tailoring the Recovered Charge in Power Diodes Using 2 MeV Electron Irradiation," Electrochemical Soc. Meeting, Paper 261RNP, 1975.

14. B.J. Baliga and E. Sun, "Lifetime Control in Power Rectifiers Using Gold, Platinum and Electron Irradiation," IEEE (1976), IEDM Technical Digest, Washington, D.C., (1978) 495.

15. J.L. Brown, U.S. Patent No. 3877997.

16. A. Jaecklin, et al., U.S. Patent No. 3943549.

DISCUSSION

(Chairman: Prof. P. Leturcq, Institut Nationale des Sciences Appliquées, Toulouse)

C. Schüler (Brown Boveri Research Center, Baden)
You said that a new contact for improved cooling is not totally hypothetical. Would you expand on that?

M. Adler (General Electric, Schenectady)
The method here was to use a structured copper with small individual copper wires packed about 90% dense and that contact the silicon. Metallurgical contact is thereby achieved without the problem of thermal expansion mismatch that occurs if you solder a block of copper directly to the silicon.

C. Schüler
Is this a method you are actually using in commercial products?

M. Adler
Not yet, but it's certainly further along than it was when these data were first produced.

R. Sittig (BBC Brown, Boveri & Co. Ltd., Baden)
What conditions are chosen for the comparison of carrier concentrations in thyristors doped with gold or platinum or electron irradiated? Are the current densities kept constant?

M. Adler
All were done at the same current density.

H. Irmler (Semikron GmbH, Nürnberg)
What materials were you thinking of with regard to diffusion bonding?

M. Adler
Direct bonding of copper to silicon.

M. Melchior (ETH, Zurich)
You have shown very impressive results from the modelling throughout the device. Can you give more information about what you can model and is documentation about this available?

M. Adler
We can do quite a bit of modelling. The model in its present form can handle both thermal as well as electrical effects. It also permits embedding the device in a

circuit. You can put a snubber around the device and then examine the switching behavior. Probably four or five years of development have gone into the model, and details are not available.

M. Melchior

But you have generic descriptions of the model?

M. Adler

Yes, I'm happy to talk to anyone about the nature of the model.

PHYSICAL LIMITATIONS AND EXPLORATORY DEVICES[*]

P. ROGGWILLER AND R. SITTIG

BBC Brown, Boveri & Co. Ltd., Baden, Switzerland

SUMMARY

Regardless of the specific type of semiconductor power device, the two stationary states "on" and "off" and the transients to high voltage dV/dt and to high current dI/dt can be identified. It is shown that between field-effect and bipolar devices a considerable difference exists in the relation of maximum forward-current density to breakdown voltage. This difference results from the characteristics of the semiconductor material, especially charge-carrier mobility and maximum field strength. The high conductivity of current bipolar devices is combined with limitations of the switching speed.

The maximum current density is usually not limited by the semiconductor but by heat removal. New packaging concepts seem to allow an improvement of surge-current capability and average on-state current by a factor of 2 to 3. The maximum breakdown voltage may also be raised further. But in this case progress is combined with an enhanced sensitivity to overload and stricter requirements on the perfection of the crystals.

The dV/dt transient depends strongly on the structure of the device and on the mode of operation. Since no principal limitation exists and the maximum ratings of present bipolar devices are relatively low, progress here will depend on the requirements of the circuit. The same holds for the dI/dt transient. The ultimate limit is several orders of magnitude away from today's general needs, and can be approached only at high cost.

A new device is presented which is controlled by electron irradiation. It exhibits high conductivity in the on-state, but the switching speed does not depend on charge-carrier lifetime. Safe operation at switching frequencies up to 200 kHz are demonstrated. This device, however, requires the integration of an electron gun and a semiconductor diode in a vacuum-tight package.

[*] Presented at the symposium by R. Sittig

1. INTRODUCTION

Progress in the development of semiconductor devices and in the further optimization of their characteristics is achieved by an accurate analysis of numerous detail problems and by elaborating applicable solutions. From time to time, however, it is worthwhile to consider things from a wider perspective in order to receive an impression of different trends, to become aware of the approach to physical limitations, ultimately to discover different solutions of technical problems, and to learn about other influences which are often connected to cost aspects. Surprisingly it often turns out that only a few years after such "fundamental" considerations, solutions are proposed which completely overcome the foreseen limits. The specification of problems and illustration of possible directions of progress seem to greatly stimulate the development of improved solutions.

The main function of semiconductor power devices is either to block a voltage or to conduct a current. They may be considered as "digital" elements exhibiting only these two states. All operating conditions in which current and voltage are simultaneously present as stationary states have to be avoided, because the resulting power losses would cause the destruction of the components in a very short time. The situation is shown schematically in Fig. 1 for a rectifier and a switch.

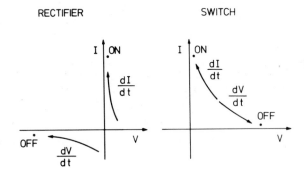

Fig. 1   Schematic current-voltage diagram for power devices showing the stationary states "on" and "off", and the transients dI/dt and dV/dt.

The goal is:
- high blocking capability at no current
- high current-carrying capability at no forward voltage drop
- no losses in the device and no restrictions on the circuitry associated with transient processes to the high-voltage state regardless of the previous current amplitude

- no losses and restrictions associated with the transient to a high-current state regardless of the previous voltage

Some aspects of the physical interdependencies between different characteristics and possible destruction mechanisms will be discussed.

2. DEPENDENCE OF ON-STATE CURRENT DENSITY ON BREAKDOWN VOLTAGE

The maximum breakdown voltage V_{BD} that can be reached by a semiconductor device depends on its "base" thickness, W_B, and doping concentration N_B

i.e.
$$V_{BD} = V_{BD} (W_B, N_B) \tag{1}$$

In the simple case of abrupt junctions there is a formula in the form of equation (1) which for silicon is

$$V_{BD} \simeq 6 \cdot 10^{13} \cdot N_B^{-3/4} , \tag{2}$$

with V_{BD} in volts for N_B in cm^{-3}.
The extension of the space-charge layer at V_{BD} is

$$W = \sqrt{\frac{2\varepsilon}{q} \frac{V_{BD}}{N_B}} \tag{3}$$

where ε is the dielectric constant and q the elementary charge. Using equation (2) we arrive at

$$W_{Bmin} = 2.3 \cdot 10^{-6} V_{BD}^{7/6} \text{ [cm]} \tag{4}$$

for the dependence of minimum base width required to withstand a breakdown voltage V_{BD}.

If the charge carriers of the device in its on-state are the majority carriers of the base, then we can calculate the minimum on-resistance.

Using
$$\rho = \frac{1}{q\mu N_B} , \tag{5}$$

and

$$R_{on} = \frac{W_{Bmin} \cdot \rho}{A}$$ (6)

where μ is the carrier mobility and A the active area in cm^2, and with $\mu \simeq 620\ cm^2\ V^{-1}\ s^{-1}$, one obtains

$$A \cdot R_{on} = 10^{-8}\ V_{BD}^{5/2}$$ (7)

The current density at a forward voltage drop of 1.5 V thus results in

$$j = 1.5 \cdot 10^8\ V_{BD}^{-5/2}$$ (8)

for majority-carrier devices as field-effect transistors, for example.

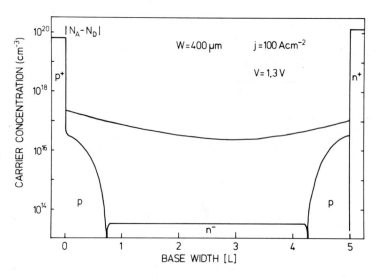

Fig. 2   Concentration of injected carriers in a typical thyristor structure. The carrier concentration profile and the voltage remain constant, if the proportion of base width to diffusion length, W/L, is kept constant and the current density is varied inversely with the base width.

The dependence $j(V_{BD})$ is correct even for other semiconductor materials, but the value of the proportionality constant depends on breakdown field strength and carrier mobility. In the case of gallium arsenide, for example, it is about a factor of ten larger.

In bipolar devices the carrier concentration in the conducting state can be increased by some orders of magnitude. These devices exhibit no ohmic behavior, but we can again derive a relationship for the current density at 1.5 V forward voltage drop by taking into account scaling laws.

Fig. 2 illustrates the doping profile of a typical thyristor and the concentration of free charge carriers in the lightly doped zones. The width is measured in multiples of a diffusion length, which means that an equivalent situation is maintained if, with a change of the width by a factor F

$$W^* = F \cdot W, \tag{9a}$$

the lifetime τ is changed according to

$$\tau^* = F^2 \tau . \tag{9b}$$

Such a variation changes the electric field

$$E^* = \frac{1}{F} E \tag{9c}$$

and the gradient of carrier concentration to

$$\frac{dn^*}{dx} = \frac{1}{F} \frac{dn}{dx} \tag{9d}$$

and therefore at constant carrier concentration and constant voltage drop across the base, V_B,

$$j^* = \frac{1}{F} j \tag{9e}$$

The junction voltage is kept constant under these conditions since the carrier concentration remains constant at the junction.

Assuming a typical base width of W = 5 L, then one obtains at 1.5 V the approximate formulae

$$j \simeq 5 \cdot 10^6 \ V_{BD}^{-7/6} \ [A \ cm^{-2}] \qquad (10)$$

and

$$\tau \simeq 1.5 \cdot 10^{-14} \ V_{BD}^{7/3} \ [s] \qquad (11)$$

These dependencies of current density on breakdown voltage for field-effect devices, equation (8), and bipolar devices, equation (10), are shown in Fig. 3. At 500 V the possible current density of the bipolars is two orders of magnitude higher and its slope is only one half that of the field-effect devices.

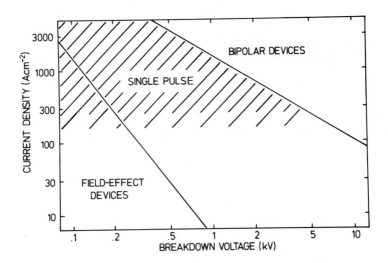

Fig. 3   Maximum current density at a forward voltage drop of 1.5 V versus break-down voltage for field-effect and bipolar devices. Due to the limitations of present-day cooling systems the hatched area can only be utilized for single pulses.

In reality the situation is not as impressive as demonstrated here since only current densities of some hundred A cm^{-2} can be tolerated during continous operation due to limited heat removal. This limitation is indicated in Fig. 3 by the hatched area. The figure thus demonstrates that up to some hundred volts, only the higher surge capability of bipolar devices might be of advantage compared to field-effect devices. At higher voltages, however, a field-effect device would need an active area which is orders of magnitude larger than that of a bipolar device for the same current.

As mentioned above the advantage of the high current density of bipolars is combined with a considerable drawback of an increasing lifetime. Equation (11) is shown in Fig. 4. While the carrier lifetime may be as short as 30 ns at 500 V, it has to be increased to 1 μs at 2 kV. As a rough estimate of the corresponding turn-off time for thyristors $t_q \sim 20$ τ is also indicated.

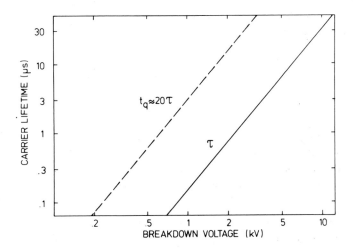

Fig. 4   Variation of carrier lifetime τ versus breakdown voltage. The base width is assumed to be 5 diffusion lengths. The curve is calculated for the minimum width. For typical conventional thyristors it has to be raised by a factor of two.

3. LIMITATIONS AT HIGH CURRENT DENSITIES

The on-state current density of high-power devices is limited by heat removal. A typical value of the thermal resistance between the silicon wafer and the water flowing through a cooler is

$$R_{th\ JW} \simeq 0.67 \text{ K cm}^2 \text{ W}^{-1} \tag{12}$$

Assuming a temperature difference of 100 K between junction and water, an average power of

$$P \simeq 150 \text{ W cm}^{-2} \tag{13}$$

can be removed from the silicon. This value limits the current density to about 100 A cm^{-2} during continuous operation. It is, however, not a physical limitation. It has been shown[2] that a dramatic improvement is possible if the silicon surfaces can be kept at constant temperature. This would decrease the thermal resistance to less than one percent of the value given in equation (12). It is of course nearly impossible to reach this ultimate physical limit. A more realistic impression of possible improvements are obtained by considering the different temperature differences which add up to 100 K. In a typical example for a present-day package these differences are:

$$
\begin{aligned}
\Delta T \text{ junction - case} &\quad \sim \quad 60 \text{ K} \\
\Delta T \text{ case - cooler} &\quad \sim \quad 10 \text{ K} \\
\Delta T \text{ cooler - water} &\quad \sim \quad 30 \text{ K}
\end{aligned}
\tag{14}
$$

If an ingenious way could be found to directly join the silicon wafer to the cooler, thereby avoiding the intermediate metal pieces and dry interfaces, then the thermal resistance could be decreased by a factor of 3. This represents an interesting improvement and therefore there are several activities in this direction.

During surge-current pulses the heat cannot be removed so that the silicon wafer and its direct surroundings are heated up. Then the temperature rises according to

$$\frac{dT}{dt} = \frac{jV}{C} \tag{15}$$

where C represents the effective thermal capacitance. The physical processes occurring under these conditions have been investigated by several authors.[3,4]

Taking into account

$$j \sim (\mu_n + \mu_p)^2 \, V^2 \tag{16}$$

where μ_n and μ_p are the electron and hole mobilities and V is the voltage across the base, which represents almost the total forward voltage drop at high-current densities and high temperatures, and

$$(\mu_n + \mu_p) \sim T^{-5/2} \ , \tag{17}$$

then one obtains

$$\frac{1}{j}\frac{dj}{dt} = g(j) \left(\frac{1}{V}\frac{dV}{dt} - \frac{5}{2}\frac{jV}{TC} \right) . \tag{18}$$

g(j) is a function which slightly decreases with increasing j from a value of 2 to about 1.3.

From equation (18) one expects a current-limiting region due to the decrease of the first term in brackets and a simultaneous increase of the second term with increasing voltage. This has been observed by Silber and Robertson.[3] There is a fast temperature rise during such a situation leading to strong increase of the intrinsic carrier concentration, n_i. When n_i approaches the concentration of free carriers in the base region, the recombination rate drops and the device becomes unstable, typically at temperatures of 400 - 500°C. A local increase of carrier density and current density occurs, causing further temperature rise and carrier generation until the device is destroyed.

The only means of avoiding this destruction mechanism is to construct a package exhibiting a large effective thermal capacitance, to keep the temperature of the device low at normal operation, or to avoid exceeding surge currents by special precautions in the circuitry.

4. LIMITATIONS AT HIGH BLOCKING VOLTAGES

Due to the strong dependence of carrier lifetime on base width as indicated by equation (9b) there is a demand to decrease W for constant breakdown voltages. This has led to the development of p-i-n structures as described in a previous paper.[5] Due to the decrease of maximum electric field with decreasing doping concentration there exists an absolute maximum of the breakdown voltage versus doping concentration at a given base width. This dependence is shown in Fig. 5. If one assumes abrupt junctions, then the maximum is obtained when the width of the lightly doped n-base corresponds to one quarter of the width of the space-charge region without an n-blocking layer at the same doping concentration. In spite of this reduction in width the voltage still amounts to 7/16 of the breakdown voltage of a device without a blocking layer.

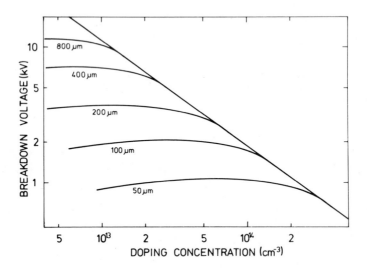

Fig. 5   Breakdown voltage versus doping concentration at different widths of the lightly doped region. p^+ n^- n^+ structures with abrupt junctions are assumed.

It follows from Fig. 5, however, that a device constructed according to these guidelines may not be subjected to high reverse-current densities. Then the free charge carriers would compensate the space charge, the breakdown voltage is reduced and the device becomes unstable.

The characteristics of a p-i-n diode at high voltages and high-current densities have been investigated in more detail by Egawa.[6] At high electric fields the current density is proportional to the total concentration of free charge carriers, because electrons and holes move with the scattering-limited velocity. The free charge carriers change the space charge and thereby the electric-field distribution may be completely changed. We have recalculated stationary states using more realistic expressions for the field dependence of the ionization coefficients α_n, α_p. The resulting distribution of the electric field at different free-carrier densities is presented in Fig. 6.

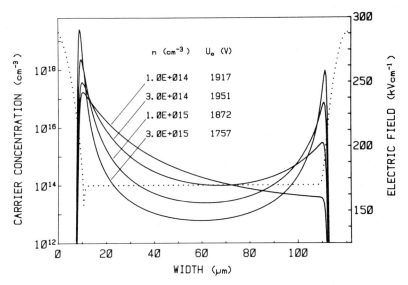

Fig. 6   Electric field versus width for different carrier densities. Strong avalanche multiplication occurs. The resulting current density is proportional to carrier density times 1.6×10^{-12}.

The field strength at the center is reduced, while a second field maximum occurs at the n⁻-n junction. With increasing current density both maxima increase further. The corresponding concentration of free electrons is shown in Fig. 7. Egawa[6] has demonstrated that in a current-voltage diagram this effect causes a swing from a positive to a negative resistance characteristic. When this point is reached, high frequency TRAPATT oscillations[7] are usually observed in realistic circuits, followed by the immediate destruction of the device. The absolute voltage maximum of the characteristic compared to the breakdown voltage at zero current decreases with decreasing doping concentration in the lightly doped layer.

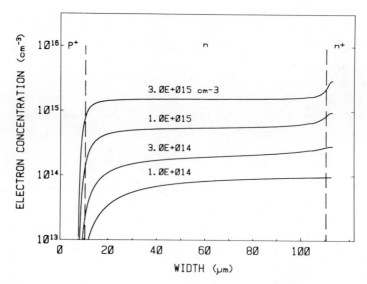

Fig. 7   Concentration of free electrons corresponding to the field distributions of
          Fig. 6.

This failure mechanism can occur at high reverse overload or during fast turn-off, when there are still high concentrations of charge carriers present. It may also appear locally, caused by very small defects at low-current levels. As the effect becomes more probable with increasing breakdown voltage and decreasing base doping concentration, silicon quality may again become a limiting factor in the development of 10-kV devices.

5. PROCESSES OCCURRING DURING A FAST RISE OF
BLOCKING VOLTAGE dV/dt

The simplest case occurs for a diode structure with no injected carriers. Due to the rising blocking voltage there is a capacitive current density

$$j = q \ C(V) \ \frac{dV}{dt} \tag{19}$$

where $C(V)$ represents the capacity of the p-n junction. If we neglect generation, no carriers flow through the region of maximum field strength and consequently no multiplication process occurs. The current at the metallurgical junction is simply the displacement current

$$j = \varepsilon \frac{dE}{dt} \quad . \tag{20}$$

If the breakdown voltage is not reached this is a safe process independent of the dV/dt.

During a voltage rise up to V_{BD} an energy

$$E_c = \int j V dt = \frac{2}{3} C(V_{BD}) \, V_{BD}^2 \tag{21}$$

is stored in the junction capacitance. The losses occurring in the device during a charge-discharge cycle depend strongly on the conditions applied by the circuit. Without a specification of the device type and the circuit, E_c may be considered to indicate the order of magnitude of switching losses due to majority carriers alone.

Using equation (2), the loss for one pulse is found to be

$$E_c \simeq \frac{10^{12}}{N_B} \; Ws \; cm^{-2} \tag{22}$$

For a high-voltage device $(N_B = 5 \cdot 10^{13} \; cm^{-3}, \; f = 50 \; Hz)$ this results in minimum losses of 1 W cm^{-2}.

For a fast device at low voltages $(N_B = 8 \cdot 10^{14} \; cm^{-3}, \; f = 100 \; kHz)$ one obtains 125 W cm^{-2}, which is nearly the same as the average power that can be removed by the normally used package according to equation (13).

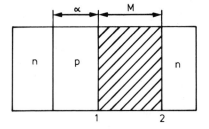

$$I_n(1) = \alpha I$$
$$I_n(2) = M I_n(1)$$
$$M \alpha I = I$$

INSTABLE IF $M\alpha > 1$

Fig. 8   Sketch of the mutual interaction of avalanche multiplication and transistor amplification if current flows across two p-n junctions at high voltage.

If there are still injected carriers present during the rise of blocking voltage, then the total losses can be considerably larger, and multiplication of carriers has to be taken into account. Nevertheless it is a stable process at all values of dV/dt for a simple diode structure if V_{BD} is not reached.

The situation is totally different if the current has to flow across a second p-n junction as in transistors and thyristors. This is indicated in Fig. 8. The second junction is forward biased and forms the emitter of a transistor. The criterion of stability is now

$$M\alpha < 1 \qquad\qquad (23)$$

where M is the multiplication factor in the space-charge layer and α the current gain of the transistor (which depends on j and time during a dV/dt pulse).

Using the approximate formula[1]

$$M(V) \simeq \frac{1}{1-(\frac{V}{V_{BD}})^3} \qquad\qquad (24)$$

one finds that a device becomes unstable at $V = V_{BD}/2$ if $\alpha = 0.86$. This is one mechanism of the dangerous second breakdown of power transistors and gate turn-off thyristors.[8] Device failure by this process can only be avoided by restricting the dV/dt or by short circuiting the hole current, so that it cannot drive the emitter junction in high injection. For transistors, narrow emitter stripes are used and the hole current can be controlled via the base contact. For thyristors the "emitter shorts" are well established. They exhibit the desirable characteristic of short circuiting a large area around them at medium current densities with a dV/dt pulse, but only slightly degrade the emitter efficiency at high-current densities in the normal on-state. This is shown in Fig. 9. The p-base is connected to the cathode contact over a circular spot of radius r. In the simplest case one can assume a constant current density entering the p-base. Around the short the current will flow laterally to the contact. The ohmic voltage drop of this current causes a forward bias at the n^+ - emitter. At distance R_ℓ, where the bias amounts to about 0.6 V, the conditions of high injection are fullfilled. On the right-hand side of the figure the dependence of R_ℓ/r on current density j is shown. The shape of this curve is constant, but it can be shifted along the j-axis by varying the p-base sheet resistance or the short radius. Between 0.1 A cm^{-2}, which is a typical value during dV/dt pulses and 50 A cm^{-2}, R_ℓ/r is reduced by a factor of 10. This means that if the total cathode area is short circuited by a dV/dt pulse, then only 1 % of the area is inactive at normal on conditions.

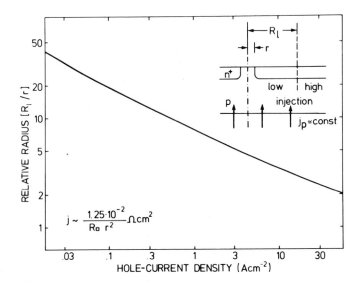

Fig. 9   The relative radius of the low-injection area around an emitter short as a function of hole-current density.

6. PROCESSES LIMITING THE RATE OF CURRENT RISE, dI/dt

Silicon offers the opportunity of creating high densities of charge carriers in the bulk within some picoseconds. This can be achieved by irradiating the crystal with a high-power light pulse of a neodymium-YAG laser. This mode of turn-on has been investigated by several authors.[9,10] With light pulses of about 10 ns duration, carrier densities up to 10^{20} cm^{-3} were generated in the base region. This carrier concentration is nearly three orders of magnitude higher than that at normal forward conduction. Correspondingly high peak-current densities of more than 50,000 A cm^{-2} were observed. Although the circuit was of extremely low inductance, dI/dt was circuit-limited and amounted to 750 kA $(\mu s)^{-1}$. Turn-on losses are extremly low under these conditions. This method can overcome all practical dI/dt limitations, but the requirement of using a neodymium-YAG laser restricts its use in general applications.

Another way of very fast generation of charge carriers was investigated by Grekhov et al.[10] As mentioned earlier (Section 6) there are nearly no charge carriers in the region of maximum electric field at high rates of voltage rise starting from zero current. Using a dV/dt of about 10^6 V $(\mu s)^{-1}$, a retardation of the onset of multiplication was observed. The stationary breakdown voltage could be considerably exceeded. After a delay of about 2 ns this causes an extreme pulse of carrier generation and a dI/dt of about 70,000 A $(\mu s)^{-1}$. This method again demonstrates that the ultimate physical limitations of the rate-of-current rise are far beyond today's requirements.

In practical systems we have to consider the injection of charge carriers from the emitter junction. The problem then is that the carriers have to move across the entire base width to be stored at the opposite emitter, which will then increase its injection of carriers of the other kind. During the first phase of turn-on the current density is nearly independent of the applied voltage but is limited by the already stored charge, Q. In a first approximation one expects

$$j = k\, Q \tag{25}$$

where k is a constant of proportionality. If one considers a one-dimensional thyristor structure and neglects the gate current, the rate of increase of Q is given by

$$\frac{dQ}{dt} = j\,(\gamma_n + \gamma_p - 1) - \frac{Q}{\tau} \tag{26}$$

where γ_n and γ_p are the efficiencies of the cathode and anode emitter, which are thought to include the effect of the emitter shorts. From equations (25) and (26) one obtains an exponential rate of rise of current j and charge Q with a time constant

$$t_r = \frac{1}{k\,(\gamma_n + \gamma_p - 1) - \frac{1}{\tau}}\,. \tag{27}$$

The time constant becomes short when no charge carriers leave the base regions ($\gamma_n = \gamma_p = 1$), when recombination may be neglected ($\tau \gg t_r$) and when the constant k becomes large. k can be interpreted as the efficiency of Q to contribute to the current density j. It depends on the doping profile, the width of the base layers, the carrier mobilities and the extension of the space-charge region. Obviously k is larger in the case of high injection, when the electric field can support the current density, than at low injection. A quantitative analysis of the time constant t_r for some special cases is discussed in the textbook of Gerlach.[12]

Although this is far from a well-founded description of the complicated turn-on process, it may help in judging the qualitative influence of the different parameters of a thyristor structure. Moreover it has been suggested[13] that the time constant t_r is of major importance for the velocity of plasma spreading.

An important aspect of the dI/dt failure of thyristors is that the circuit reacts to total current, or its rate of increase, whereas the device is subjected to the local current density, which may be three orders of magnitude higher than during the stationary on-state. To avoid this destruction mechanism one, therefore, tries to ensure that a large enough area is turned on immediately. This concept is well established and has been reported elsewhere.[5]

7. ELECTRON-BOMBARDED SEMICONDUCTOR DEVICES

In accordance with the above considerations, EBS-devices appear to be very promising. These structures have been investigated and developed as very high-frequency amplifiers.[14,15]

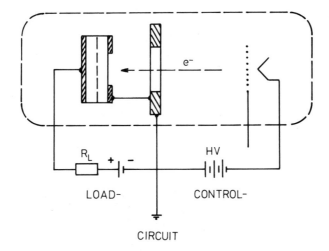

Fig. 10   Schematic diagram of an electron-bombarded semiconductor device. A high-voltage electron beam generates a current in the reverse-biased diode with a multiplication factor of 2000-3000.

The principle is shown in Fig. 10. This hybrid device consists of a vacuum tube and a semiconductor diode. The beam of 10-20 keV electrons can be controlled by the grid voltage. Each electron impinging on the semiconductor diode creates about 2000-3000 electron-hole pairs. They are separated due to the internal fields in the diode structure and a reverse current through it is generated which can be about 2000 to 3000 times as large as the current of the electron beam.

In the case of high-frequency amplifiers the doping profile of the diode is chosen, so that the space-charge region extends to very close to the surface even when only ten percent of the breakdown voltage is applied. Such operating conditions are maintained to ensure that practically all electron-hole pairs contribute to the current, and that all charge carriers move with their scattering-limited velocity of 10^7 cm s^{-1}. This allows the operation of the device up to frequencies of 10 GHz.

For a power switch, high voltage in the on-state must be avoided. But even with no voltage drop across the diode a high reverse current can be generated. The behavior is similar to a solar cell at high irradiation. Measurements of this characteristic are shown in Fig. 11. The low-current densities are chosen to limit the temperature rise due to electron bombardment for these d.c. measurements. At low currents the diode can even deliver power to the circuit. There can be a high-current density at zero voltage drop.

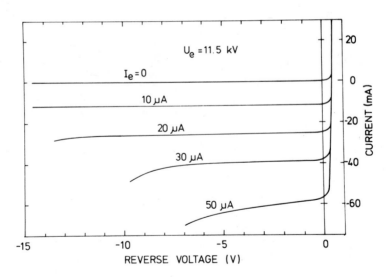

Fig. 11   Reverse-diode characteristic with electron irradiation of 11.5 kV and different beam currents.

The stationary current is limited either by the generation rate, when it does not depend on the applied voltage, or is limited by the load resistance. To avoid unnecessary losses the irradiation intensity has to be matched to the load. It has to be large enough to ensure that no rest voltage occurs in the on-state but must not be too large in order to keep the irradiation losses as small as possible.

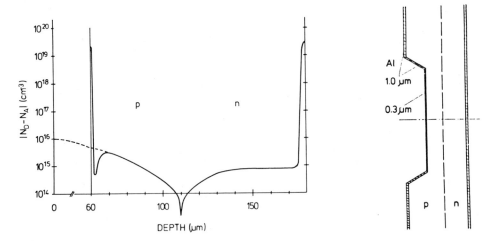

Fig. 12   Doping profile and surface contour of the diode investigated.

The structure of the diode which we have investigated is illustrated in Fig 12. A deep aluminum profile has been diffused into a thin wafer containing a concentration of phosphorus atoms of $5 \cdot 10^{14}$ cm^{-3}. Sixty microns of the p-profile have been etched away over a 3-mm diameter to produce the active area. This allows the use of wafers of manageable thickness and the application of the standard technique of positive bevelling at the outer edge of the device. The diode structure is completed by phosphorus diffusion from the backside and boron diffusion with a penetration depth of about 1 μm from the front. The diodes exhibit a breakdown voltage of 1000 V. At this voltage the space-charge layer extends 18 μm into the p-zone and 38 μm into the n-substrate. In fully optimized devices of the same thickness a breakdown voltage of about 2 kV would be possible. An aluminum layer of 0.3 μm thickness is evaporated onto the bottom of the etched groove and 1 μm-thick aluminum layers are used as contacts on the rest of the front and on the backside. This device is mounted on a water-cooled holder in an experimental vacuum arrangement as shown in Fig. 13.

The electrons are emitted from a hot tantalum cathode in the experimental arrangement. The electron beam is limited by a diaphragm, so that only the active area of the device is irradiated. The control circuit permits the beam to be switched with a rise time of 120 ns.

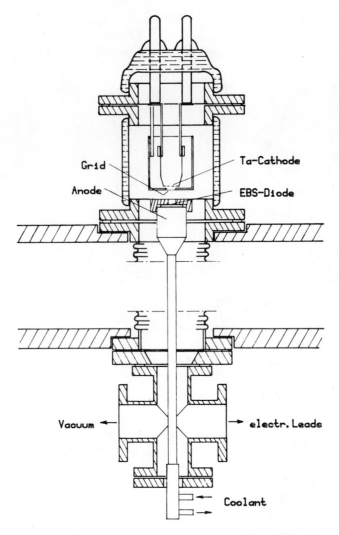

Fig. 13   Experimental setup used in the investigation of the EBS-device.

At an energy of 15 keV about one third of the power of the electron beam is absorbed in the aluminum layer while the rest is transferred to the silicon. By reducing the thickness of the aluminum layer to 0.1 μm only half the losses would occur. The maximum carrier-generation rate occurs in a depth of 0.8 μm. Each electron generates 2100 electron-hole pairs. The electron-hole pairs can be separated by the electric field of the p^+-profile or of the space-charge layer so long as it extends nearly throughout the lightly doped p-region. An increase in current will reduce the applied voltage if an ohmic load is considered. The extension of the space-charge region shrinks and more carriers have to be stored in the p-base to allow a further increase of current.

After only a short delay, both lightly doped zones are swamped by injected carriers and the current reaches its saturation level. A calculation of the pulse shape and the amount of stored charge is presented in Fig. 14.

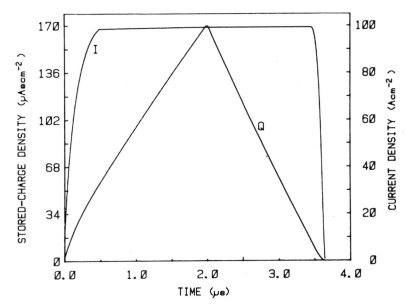

Fig. 14   Calculated time dependence of the diode current and the stored charge under the assumption of a rectangular generation pulse of 190 A cm^{-2} during 2 μs at 400 V and a load of 4 Ω.

At the end of the irradiation pulse the current will continue to flow maintained by the stored charges. Since the p-n junction is nearly at the center of the diode, and the p-profile exhibits a very small gradient there, all stored carriers have to be removed before the diode can return to the blocking state. This occurs with an abrupt decrease in current and a sharp increase in voltage.

Fig. 15 shows the diode current and the voltage pulses which were obtained with an electron-beam current of 1.43 mA at 15 kV. The load circuit consisted of a 200-Ω resistance and negligible inductance. Therefore current and voltage exhibited precisely the same waveforms. As expected, there is a slight delay of current rise due to the required storage charge. The amount of the retardation and the corresponding turn-on losses can be influenced by an overdrive of the beam current during the rise time. But even at a constant irradiation pulse, which just suffices to drive the diode into full saturation, only 400 ns were needed to reach the final current density of

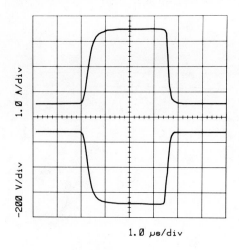

1. 0 μs/div

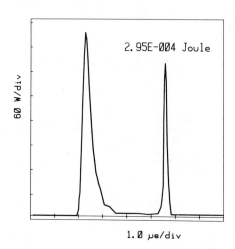

1. 0 μs/div

Fig. 15   Upper part: measured current and voltage pulse shapes for the EBS-device
with an electron-beam current of 1.43 mA. Lower part: resulting switching
losses.

43 A cm^{-2}. In contrast to thyristors a high dI/dt does not represent a critical phase of operation, since the total active area is irradiated simultaneously and no lateral plasma spreading has to take place. The diode current stayed constant for about 300 ns after termination of electron irradiation and then dropped sharply to zero within about 200 ns, corresponding to the RC-time constant of the load resistance and junction capacitance.

The resulting switching losses are presented in the lower part of Fig. 15. For an active area of 1 cm^2 they amount to $3 \cdot 10^{-3}$ J cm^{-2} for turn-on and $1 \cdot 10^{-3}$ J cm^{-2} for turn-off. For an estimate of the total losses we have to take into account the irradiation losses, P_i, while the stationary on-state and off-state losses may be neglected. With a duty cycle of t_{on}/T and the applied current density of 43 A cm^{-2} one obtains

$$P_i = 300 \ t_{on}/T \quad [W \ cm^{-2}] \qquad (26)$$

The switching frequency is limited by heat dissipation and not by charging or discharging time constants. In fact it was proven that the device could operate at 200 kHz at low-current density.

From the point of view of physical limitations this device exhibits several advantages:
- It is a bipolar device which may be used at high-current densities, even at breakdown voltages of more than 1 kV.
- Charge carriers are generated homogeneously across the total active area by external irradiation. No plasma spreading has to take place.
- The extraction of charge carriers is achieved by an electric field at very low injected carrier densities. As there is only one p-n junction this phase of operation is completely safe, requiring only precautions to avoid overvoltages.

The drawbacks arise from the hybrid structure of the device and from the necessity to use a vacuum tube. Therefore a prerequisite for the application of EBS-devices is the development of cold electron-emitting cathodes of unlimited lifetime. According to a recent paper[16] this could become possible within the near future. The use of EBS-devices may then be a realistic possibility.

REFERENCES

1. S.M. Sze, Physics of Semiconductor Devices, New York: John Wiley & Sons, 1969.

2. M.S. Adler and H.H. Glascock, "Investigations of the Surge Characteristics of Power Rectifiers and Thyristors in Large-Area Press Packages," IEEE Trans. Electron Devices, ED-26 (1979) 1085-1096.

3. D. Silber and M.J. Robertson, "Thermal Effects on the Forward Characteristics of Silicon p-i-n Diodes at High Pulse Currents," Solid-State Electronics, 16, Part II (1973) 1337-1346.

4. Y.C. Kao and P.L. Hower, "The Surge Capability of High Voltage Rectifiers," IEDM Technical Digest, Washington (1978) 568.

5. P. De Bruyne, J. Vitins and R. Sittig, "Reverse-Conducting Thyristors," present volume.

6. H. Egawa, "Avalanche Characteristics and Failure Mechanism of High Voltage Diodes," IEEE Trans. Electron Devices, ED-13 (1966) 754.

7. B.C. DeLoach and D.L. Scharfetter, "Device Physics of TRAPATT Oscillators," IEEE Trans. Electron Devices, ED-17 (1970) 9-21.

8. D.L. Blackburn and D.W. Berning, "An Experimental Study of Reverse-Bias Second Breakdown," IEDM Technical Digest, Washington (1980) 297.

9. J.R. Davis and J.S. Roberts, "Ultra-Fast, High-Power Laser-Activated Switches," PESC-Record, Cleveland (1975) 272.

10. C. Defois, "Contribution à l'étude de la fermeture des thyristors de grande puissance: Commande électrique et commande optique," thesis, presented at L'Institut National des Sciences Appliquées de Toulouse, France, 1978.

11. I.V. Grekhov, A.F. Kardo-Sysoev, L.S. Kostina and S.V. Shenderey, "High Power Subnanosecond Switch," IEDM Technical Digest, Washington (1980) 662.

12. W. Gerlach, Thyristoren. Halbleiter-Elektronik, vol. 12, Berlin: Springer-Verlag, 1979.

13. Power Devices Workshop, working group 4, National Bureau of Standards, Washington, D.C. 1980.

14. C.B. Norris "Optimum Design of Electron Beam-Semiconductor Linear Low-Pass Amplifiers," IEEE Trans. Electron Devices, ED-20 (1973) Part I, 447, Part II, 827.

15. A. Silzars, D.J. Bates, and A. Ballonoff, "Electron Bombarded Semiconductor Devices," Proceedings of the IEEE, 62, Part II (1974) 1119-1158.

16. J.K. Cochran, A.T. Chapman, R.K. Feeney and D.N. Hill "Review of Field Emitter Array Cathodes," IEDM Technical Digest, Washington (1980) 462.

DISCUSSION
(Chairman: Prof. P. Leturcq, Institut Nationale des Sciences Appliquées, Toulouse)

D. Silber (AEG-Telefunken, Frankfurt)

Which emitter constants did you use in the calculation of the current density in the thyristor structure?

R. Sittig (BBC Brown, Boveri & Co. Ltd., Baden)

We did an exact calculation of an emitter layer doped to 10^{19} cm^{-3} with an abrupt junction. This results in an emitter efficiency of nearly 1.

D. Silber

Did you try to make any approximation for the limitations of a current filament flowing with high density in a high-field region? I think one has to take into account something like carrier repulsion and out diffusion.

R. Sittig

One has to take into account all carrier scattering mechanisms, but I did not try to make an estimate. I would assume that an average diameter of some microns is possible for a filament of 100 microns length. This would mean that the cross section of a filament amounts to only 10^{-7} cm^2.

JUNCTION FIELD-EFFECT DEVICES

JUN-ICHI NISHIZAWA
Research Institute of Electrical Communication,
Tohoku University, Sendai, Japan

1. INTRODUCTION

The first field-effect devices were proposed by J.E. Lilienfeld[1] in 1926 and O. Heil[2] in 1935 corresponding to a Schottky-gate device and a MOSFET, respectively. The principle of the junction field-effect transistor (JFET) was discovered by Shockley[3] in 1952 and realized as a practical device by Dacey and Ross[4] in 1953. Some years before, in 1950, Nishizawa and Watanabe[5] had applied for a patent for a similar device, which they called the "electrostatic induction transistor" (SIT). The feasibility of this device, however, could be demonstrated only in 1970 by Nishizawa, Terasaki and Shibata.[6] The fundamental idea of the SIT in 1950 was to control a resistance by means of carrier injection into a high resistivity layer using electrostatic induction. Subsequently the term "field effect" was introduced for the same phenomenon and has become widely accepted, leading to some confusion in terminology.

2. OPERATIONAL PRINCIPLE

2.1 Comparison between SIT and JFET

The main differences between the SIT and JFET will be discussed with aid of Figs. 1 and 2. Fig. 1 shows the SIT in its original form[5] from 1950. At the center of a piece of low-resistivity n-type semiconductor material is a region of high resistivity of the same type. The potential of this region can be controlled by the voltage applied to the narrowly spaced gate structure. The potential distributions in the blocking and conducting states are sketched in the lower part of the figure. Fig. 2 shows a schematic representation of a JFET. There are two gate regions on opposite sides of a channel which, in contrast to the SIT, extend over a longer distance in the direction of current flow. This structure can be more easily produced but exhibits a higher resistance, R_S, in the nondepleted part of the channel towards the source.

If an n-channel device is considered, then a negative gate voltage V_G generates a space-charge layer at the gate-channel junction. Thus V_G determines the maximum neutral channel width which is available for current flow, i.e. it determines the chan-

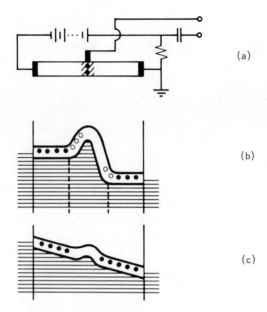

Fig. 1   The original "transistor" proposed by Watanabe and Nishizawa[5] static induction (SIT). (a) SIT design SIT and corresponding circuit, (b) energy-band
diagram when gate is biased in reverse direction, (c) energy diagram when
gate is biased in forward direction.

nel resistance. Application of a positive drain voltage V_D has to be considered with
respect to both the other electrodes. With respect to the gate the voltage difference
is increased causing a greater extension of the space-charge layer. With respect to
the source the drain voltage gives rise to a drain current I_D which is combined with
an ohmic voltage drop. This means that the potential difference between channel and
gate increases towards the drain, leading to a wider extension of the space-charge
layer into the channel in that direction. Finally the neutral channel is totally pinched
off at the drain side. At this point the I_D versus V_D characteristic starts to saturate.
Any additional drain voltage will drop across the space-charge layer between gate and
drain and cannot be utilized to increase the drain current. The ohmic voltage drop
within the neutral part of the channel, and thus the drain current, stays constant
independent of any further increase in V_D.

The SIT can be described as a JFET with a very short channel length. When the
channel length becomes shorter than the channel width the characteristics change considerably, and almost no influence of the drain current on the pinch-off is observable.

The potential distribution is sketched in Fig. 3. At moderate gate voltages (Fig. 3a) the space-charge region extends only sligthly into the channel. This corresponds to the linear I_D versus V_D characteristic, but almost no saturation is observable. The ohmic voltage drop along the short channel is adequate to complete the pinch-off only at very high drain current densities.

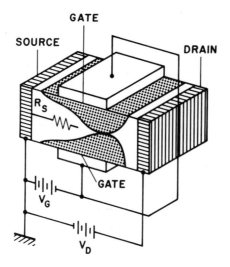

Fig. 2   Schematic representation of the operation mechanism of a junction field-effect transistor. R_S is the series resistance in the channel.

At higher gate voltages (Fig. 3b) the channel is pinched off by V_G alone. Between the gate areas a short potential barrier builds up in the form of a saddle, which reduces the drain current to very low values. The minimum barrier height occurs at the saddle point, which we refer to as the "intrinsic gate point", G'. As for the JFET, an increase of V_D for this situation contributes mainly to an extension of the space-charge layer between gate and drain. There is, however, considerable influence on the drain current.

For an exact treatment of the characteristics, the diffusion potential ϕ_{bi} between gate and channel has to be taken into account. Using $V_{G'}$ for the potential induced at the intrinsic gate due to V_G and V_D, and introducing the effective thickness of the potential barrier, $W_{G'}$, the drain-current density at G' is given by

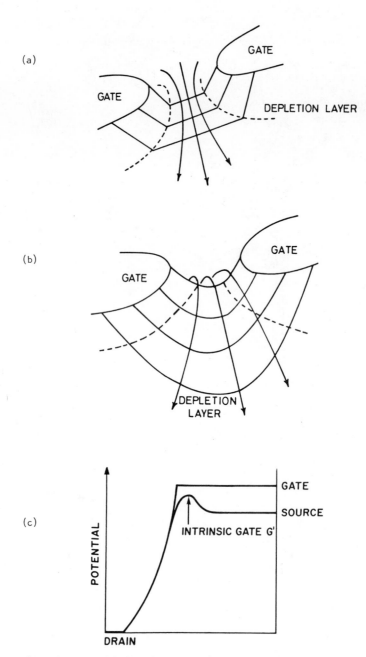

Fig. 3   Schematic potential distribution of the SIT near the gate region with arrows
          indicating direction of the current flow from the source to the drain, (a)
          linear mode, (b) exponential mode, (c) potential distribution profile of the
          SIT.

$$j_D(G') = qn_S \frac{D_n}{W_{G'}} \exp\left(-\frac{\phi_{bi} - qV_{G'}}{kT}\right) \tag{1}$$

where n_S is the carrier density in the source region, D_n the diffusion coefficient of charge carriers (electrons in this case), k the Boltzmann constant, and T the absolute temperature.

If the effective barrier thickness becomes very thin, then the diffusion velocity of the carriers crossing the barrier would be faster than the thermal velocity

i.e. $$\frac{D_n}{W_{G'}} > \sqrt{\frac{kT}{2\pi m^*}} \tag{2}$$

where m^* is the effective mass of the carriers. In this case the device was termed an "ideal device" or a "thermionic-emission device",[7] while the term "ballistic device"[8] is mainly used now. Experimental confirmation of the thermionic-emission characteristic has been published recently.[9]

In actual SIT-structures, condition (2) is not fulfilled and instead of equation (1)

$$j_D(G') = qn_S \sqrt{\frac{kT}{2\pi m^*}} \exp\left(-\frac{\phi_{bi} - qV_{G'}}{kT}\right) \tag{3}$$

should be used. The total drain-current contribution of one channel is obtained by integration of (3) across the total area between neighboring gates.

According to equation (3) the drain current varies exponentially with barrier height $V_{G'}$ and it will be shown that $V_{G'}$ depends linearly on V_D. In contrast to the JFET drain current in an SIT does not stay constant after the channel has been pinched off but depends exponentially on V_D. The difference is that in a JFET a slight reduction of the barrier height due to V_D results in a slight increase in I_D and, in turn, to a small shift of the pinch-off point towards the source. This takes place in the transaction region of the characteristic between the linear I_D-V_D dependence and constant I_D. A corresponding shift of the pinch-off point is almost impossible in an SIT due to the geometrical arrangement.

2.2 Comparison of SIT's and Bipolar Transistors

The potential distribution of an SIT along the current path, as shown in Fig. 3c, is similar to the potential in a bipolar transistor, BPT, from the emitter through the

base to the collector. Therefore it is worthwhile to compare these two types of devices. In the case of a BPT the barrier height is controlled by the total charge in the
base layer, and this has to be adjusted by a lateral current flow through the base
resistance. Accordingly the base resistance should be as low as possible, but at the
same time a highly doped base as a thick base is inconsistent with good performance
of the BPT with respect to injection efficiency, low emitter-base capacitance and high
cut-off frequency.

In the case of the SIT, the distributed gate structure corresponds to the base of the
BPT. The doping concentration in the gate area can be chosen about three orders of
magnitude higher and a correspondingly very thin "base thickness" can be used. The
switching characteristics of an SIT in some respects resemble a punching-through
BPT, and extremely high cut-off frequencies can be achieved. Of course there is no
minority carrier-current flow through the highly doped gate regions. The loss of
active device cross section is one sacrifice which has to be made.

The barrier height is controlled by electrostatic induction, which avoids the drawback
of the base resistance but requires a minimum distance between neighboring gates.
The cell size is limited by the thickness of the depletion layer at the design gate
voltage and amounts to about 10 μm for a practical SIT. In a BPT, the cell size is
limited by the voltage drop along the base resistance[10] and is usually an order of
magnitude larger, but has to be chosen much smaller for high-frequency applications.

3. NORMAL-MODE SIT

As mentioned above, the geometrical arrangement of the gate structure yields lower
channel resistances and gate capacitances and correspondingly shorter time constants
for SITs compared to FETs. The lower source resistance lowers the power gain, however, and should therefore be adjusted to the load to obtain optimum performance of
the device. The equivalent circuit of the normal-mode SIT is shown in Fig. 4, where
Fig. 4a represents the voltage source and Fig. 4b the current source. The optimum
source resistance is obtained when

$$R_S \cong \frac{R_D(\mu+1)\pm\sqrt{R_L^2(\mu+1)^2-\omega^2 C_{GS}^2(R_D^2-R_L^2)^2}}{(\mu+1)^2+\omega^2 C_{GS}^2(R_D^2-R_L^2)} \tag{4}$$

Here R_D is the drain resistance, R_L the load resistance, C_{GS} the gate-to-source capacitance and ω the operating angular frequency. The voltage-amplification factor μ is
defined as

$$\mu = -\left.\frac{\partial V_D}{\partial V_G}\right|\ I_D = \text{const.} \tag{5}$$

In normal mode operation of an SIT or FET the gate is reverse biased. Part of the total potential difference is due to the diffusion potential, ϕ_{bi}, between the gate and

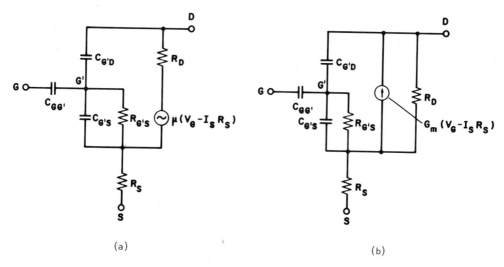

(a)   (b)

Fig. 4   Equivalent circuit of the SIT. (a) voltage-source representation, (b) current-source representation.

the channel. The extent of the depletion layer, d, from the gate junction into the channel can be calculated from

$$d \cong \sqrt{\frac{2\varepsilon}{qN_C}}\ (V_G + \frac{\phi}{q})\ , \tag{6}$$

where N_C is the doping concentration of the channel and ε the dielectric constant of the semiconductor material. When the total channel width, $2\ W_C$, is chosen according to

$$W_C = \frac{1}{q}\sqrt{\frac{2\varepsilon}{N_C}\ \phi}\ , \tag{7}$$

then the channel is pinched off due to the diffusion potential alone with no application of gate voltage. The extent of the depletion layer due to ϕ versus channel doping concentration is shown in Fig. 5. It can be seen that for a doping concentration of 10^{13} cm^{-3} a total channel width $2\ W_C = 20\ \mu$m is pinched off without an externally applied gate voltage.

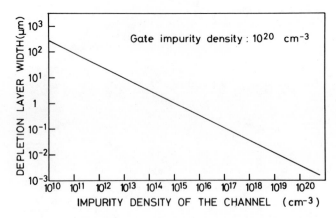

Fig. 5   The depletion-layer width as a function of the impurity density of the channel.

The output characteristics of such an SIT which has just been pinched off are shown
in Fig. 6 on a linear scale. At zero gate voltage the I_D-V_D characteristic is still
almost linear, whilst at higher gate voltages I_D first rises exponentially and then
changes over to a linear dependence with a slope that decreases only slightly with increasing V_G.

If the channel is not yet pinched off due to the diffusion potential alone, then the I_D-
V_D characteristic starts at zero gate voltage from a constant resistance characteristic
as shown in Fig. 7. The increase in channel resistivity with temperature leads to a
corresponding decrease in the drain current. At a reverse-gate bias of more than 1 V
the characteristics again exhibit the exponential dependency. The increase of thermionic-emission current with temperature is then dominant and leads to a reversal of the
temperature dependence in the low-current region.

A theoretical prediction of the I_D-V_D characteristics of the SIT at a pinch-off condition has to be based on a two-dimensional calculation of the potential distribution in
the channel between the source and the gate contact. A rough approximation can be
obtained by considering the potential height at the intrinsic gate point

$$V_{G'} = V_{G'} \ (V_G \quad V_D) \tag{8}$$

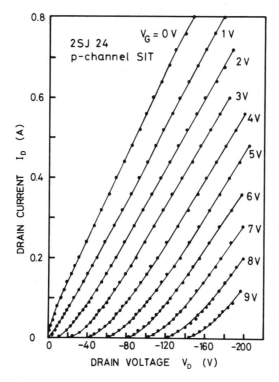

Fig. 6   Drain current vs. drain voltage of 2 SJ 24 (p-channel SIT).

Taking into account first order terms only gives

$$V_{G'} = \frac{\partial V_{G'}}{\partial V_G} V_G + \frac{\partial V_{G'}}{\partial V_D} V_D \qquad (9)$$

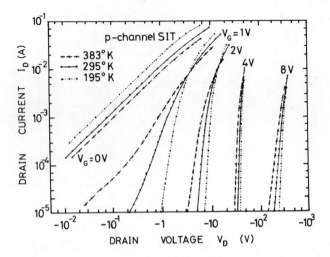

Fig. 7   I-V characteristics of the SIT for three temperatures in log-log coordinates.

The first derivative is called the effective partition ratio, η,

$$\eta = \frac{\partial V_{G'}}{\partial V_G} \tag{10}$$

The second term can be transformed into

$$\frac{\partial V_{G'}}{\partial V_D} = \frac{\partial V_{G'}}{\partial V_G} \left(-\frac{\partial V_G}{\partial V_D} \right|_{I_D = const.}) = \frac{\eta}{\mu}$$

where the voltage-amplification factor, μ, from equation (5) has been introduced. In some cases the constant drain current at which the derivative is evaluated need not be specified since μ is almost independent of I_D as shown in Fig. 8.

Inserting (10) and (11) into equation (9) leads to

$$V_{G'} = \eta V_G + \frac{\eta}{\mu} V_D \quad , \tag{12}$$

for the intrinsic gate potential. This expression together with equation (3) results directly in an exponential dependence between I_D and V_D.

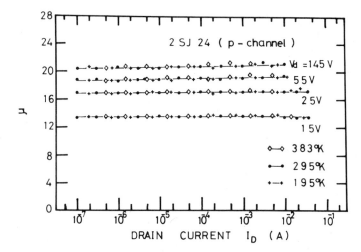

Fig. 8   Voltage amplification (μ) factor vs. the drain current in 2 SJ 24 for three different temperatures, with drain voltage as parameter.

Beyond pinch-off, approximations for η and μ can be derived using simplifying assumptions for the size of the gate area and the contour of the space-charge layer. A variation of gate voltage, ΔV_G, is combined with a variation of space charge, ΔQ, at the source-side edge of the space-charge layer, which is a distance d_S from G', and with an equal amount of charge of the opposite sign, $-\Delta Q$, at the gate area about half the channel width, W_C, from G'. Correspondingly, a variation ΔV_D results in additional charges at the drain-side edge of the space-charge layer over a distance d_D and the negative variation of charge at the gate. The influences of these charge variations on the potential at G' are easily calculated if simple geometrical arrangements are assumed. One approximation for μ is

$$\mu = \frac{d_D}{d_S} + 1 \tag{13}$$

which should be approached when the extent of the gate area in the direction of current flow is small compared to W_C which can then be neglected in relation to d_S and d_D. This is the case for low channel-doping concentration and at high drain voltages.

It follows from Fig. 7 that a variation of ΔV_G of less than 1 V is enough to increase I_D by several orders of magnitude at constant V_D. Therefore d_S stays approximately constant, and equation (13) reveals that μ depends on V_D but not on I_D or temperature. As shown in Fig. 8 this is the case for a variation of drain current from 0.1 μA to 50 mA and in a temperature range from 195°K to 383°K. The variation of drain voltage between 15 V and 145 V leads to an increase in μ from about 13 to 21. The I_D-V_D characteristics of the same device are shown in Fig. 9 on a linear scale. In contrast to the voltage-amplification factor μ, the characteristics exhibit a considerable temperature dependence.

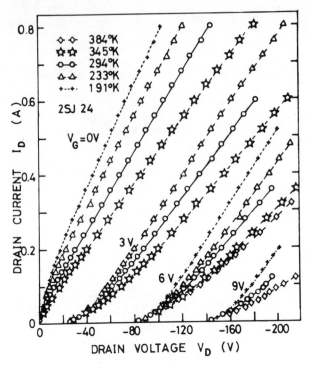

Fig. 9   I-V characteristics of the SIT (25324) at different temperatures.

An estimate of the partition ratio η has been published by Kajiwara et al.[11] who investigated a cross-type model of the SIT and obtained

$$\eta = \frac{d_S}{0.5\ W_C + d_S} \tag{14}$$

Equation (12) together with the simple equations (13) and (14) for μ and η can be used as rough design rules.

The output resistance of the SIT in a pinched-off state initially has an exponential dependence and finally approaches a linear behavior determined by the source resistance which exhibits a negative temperature coefficient as shown in Figs. 7 and 9. This guarantees stable device behavior and prevents thermal runaway. In the low-current region, the temperature coefficient is always positive in agreement with the Maxwell-Boltzmann distribution law.

From the features mentioned above it is obvious that an SIT can be produced with various structures, for example using a Schottky contact, an $n-p^+$ or a $p-n^+$ junction for the gate array or even an MIS structure. For the preparation of a very short channel SIT the use of an epitaxially grown vertical structure seems most advantageous. As an example the individual process steps for the preparation of an n-channel SIT are given below:

(1) Epitaxial growth of a very pure, high-resistivity, n-type layer from a $SiCl_4$ source on a highly doped substrate. The thickness of the epitaxial layer is determined according to the maximum drain voltage and to the voltage-amplification factor as described by equation (13).

(2) Oxidation and photoresist technology. In the case of an impurity concentration of 10^{13} cm^{-3}, the pitch of the lined gates is nearly 40 μm and the width of the window in the oxide mask is about 10 μm. Boron and germanium are diffused simultaneously using gaseous chlorides in hydrogen gas.

(3) Removal of the remaining oxide film. Epitaxial growth of a high-resistivity layer to form the low-capacitance interface between gate and source, which also increases the inverse voltage. Sometimes at the beginning of the second epitaxial growth, an additional small amount of donor impurities is added to compensate the acceptors, which diffused at the surface from the highly doped gate region. At the end of the second epitaxial deposition, the donor concentration is strongly increased to reduce the source contact resistance.

(4) The wafer is cut into single chips and a part of the second epitaxial layer is removed to open the top of the highly doped gate-contact area which is connected to the lined gates. Gold is evaporated onto both sides of the chip which is sandwiched between a pair of molybdenum plates.

A sketch of the structure of a 1 kW SIT prepared in this way is shown in Fig. 10a, while Fig. 10b represents a photograph of the chip and Fig. 10c of the packaged device. The electrical data obtained with this device are listed in Table 1, and the frequency characteristics of the voltage gain is shown in Fig. 11.

TABLE 1

Electrical Characteristics of 1.5-kW buried-gate SIT

BV_{gdo}	1,000 V ($I_g = 0.1$mA)
BV_{gso}	100 V ($I_g = 0.1$mA)
I_{dss}	20 A ($V_g = 0$V , $V_d = 10$V)
g_m	20 ℧ ($I_d = 2$A , $V_d = 50$V)
μ	15 ($I_d = 2$A , $V_d = 50$V)
C_{input}	12,000 pF ($V_g = -10$V, $V_d = 0$V)

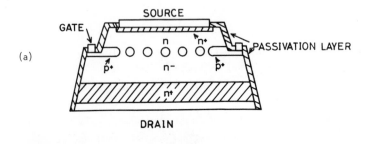

Fig. 10   1-kW power SIT with buried gate structure (a) cross section (b) top view (c) packaged.

With a pair of these SITs used in a push-pull operation for a supersonic transducer, an output power of 3 kW was obtained. The 450-kHz induction heater shown in Fig. 12 was manufactured in the same way. It seems possible that SIT's can be manufactured, which permit the output power to be increased up to about 100 kW. Then even output-stage amplifiers for broadcasting stations could be realized using these semiconductor devices.

The frequency limitation is due to the stray capacitance of the buried gate. It is expected that a step-cut structure (U-gate) can operate at higher frequencies. The maximum output power is limited by the area of the device. If the total area of a 5-inch wafer could be completely utilized, the power could reach the 1 MW range. Another limiting factor lies in the necessity of matching the impedances of input and output.

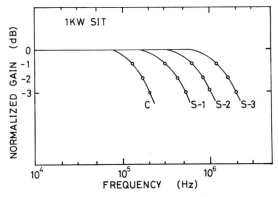

Fig. 11   Frequency characteristics of improved 1-kW SIT (low-frequency voltage gain = OdB).

Fig. 12   450-kHz push-pull amplifier using 2 chips of 1-kW SIT.

4. BIPOLAR TRANSISTOR MODE SIT (BSIT)

In applications requiring a low on-resistance, such as high-current pulse generation, the SIT can be operated with a forward voltage applied to the gate electrode. With a forward bias, minority charge carriers can be injected from the gate into the channel. For this reason this mode of operation is termed the bipolar transistor mode and a correspondingly designed SIT is named a BSIT.

With a grounded gate, the gate region is surrounded by a potential barrier due to the diffusion potential. With a narrow channel width this barrier extends throughout the channel with the lowest barrier height at the center between neighboring gates, the intrinsic gate point. The potential there decreases with increasing channel width, and therefore less drain voltage is required to drive a certain drain current.

Upon application of a forward gate voltage the diffusion voltage becomes partly compensated and thus the barrier height is reduced. Depending on whether the barrier at the intrinsic gate is merely lowered or totally removed, the drain current will exhibit either the exponential or the linear characteristic. A further increase of gate voltage totally removes the barrier. A diffusion current of minority carriers starts to flow from the gate into the channel and to the source or, at very low drain voltages, to the drain. Gate and source, as well as gate and drain, represent diode structures which are forward biased under these conditions.

When the drain voltage is increased, the gate-to-drain diode is again reverse biased, whereas the gate-to-source diode maintains the forward current. At a sufficiently high gate voltage high injection of both kinds of charge carriers from the gate and the source is reached, and the concentration of free charge carriers in the channel region may be increased by several orders of magnitude. This is combined with a corresponding increase in the drain current of the BSIT which occurs at gate voltages higher than 0.5 V, as can be seen in Fig. 13. This mode of operation seems very similar to

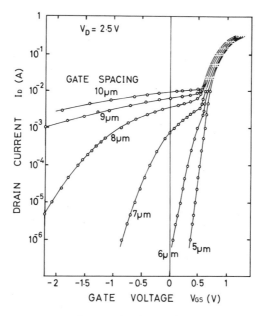

Fig. 13   Drain current gate voltage of the BSIT with six different gate spacing (V_d = 2.5 V).

that of a bipolar transistor (BPT), justifying the name bipolar transistor mode SIT (BSIT). The difference is that the neutral base layer of the BPT does not exist in a BSIT. Even at low drain voltages the electric field can extend up to the highly doped source region. This situation would correspond to base punchthrough of a BPT.

An example of the output characteristics of a BSIT is shown in Fig. 14. In the measurements the gate voltage was varied between - 1.0 V and + 0.7 V. At negative V_G

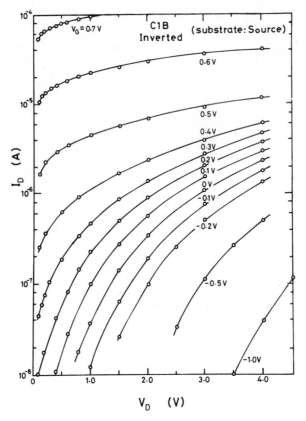

Fig. 14   Drain current drain voltage of the BSIT with gate voltages at both polarities applied.

and low drain voltage the I-V curves exhibit an exponential dependence, which changes to a linear relationship with increasing drain voltage. At a drain voltage of 4.0 V the drain current increases by almost a constant factor for each 0.1 V step of gate between - 0.2 V and + 0.4 V. For higher gate voltages this factor increases considerably indicating the transition to high injection in the gate-to-source diode.

As shown in Fig. 15, the geometrical structure of the device is rather complicated and the carrier concentration is not constant throughout the entire n^- - region in the case of high injection. The I-V curves shown in Fig. 14 nevertheless display an almost linear dependence for gate voltages of 0.5 V and 0.6 V, which appears as logarithmic on the semilogarithmic scale.

Measurements of the characteristics for the forward biasing of both the gate-to-source and the gate-to-drain diode at very low drain voltages are shown in Fig. 16. For this three-terminal arrangement the continuity equation is

$$I_D = I_S - I_G \quad ,$$ (15)

and it has to be observed that a forward current of the gate-to-drain diode represents a negative drain current. The measurements may be explained taking into account the following current contributions:

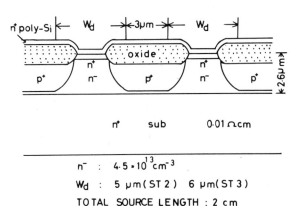

Fig. 15   Cross-sectional view of the BSIT.

(1) The gate-to-source diode forward current, which is approximately constant at constant V_G.

(2) The gate-to-drain diode forward current, which varies exponentially with $(V_G - V_D)$ and vanishes when this difference is less than about 0.5 V.

(3) The normal source-to-drain current, as shown in Fig. 14 for high drain voltages.

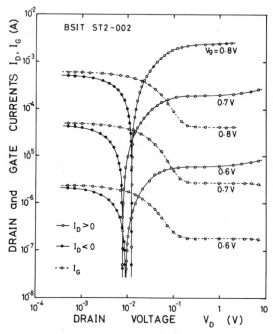

Fig. 16   I-V characteristics of the BSIT. Drain and gate current versus drain voltage with gate voltage as parameter.

At zero drain voltage the source-to-drain current vanishes, and the ratio of drain current to source current is determined by the device geometry. From the measurements it follows that about 90 % of the gate current flows towards the drain and only 10 % to the source. Assuming high emitter efficiencies for both the p^+ - gate region as well as for the n^+ - areas of source and drain, the total gate current is a hole current. Since all these holes cannot flow into the n^+ - areas, a total charge

$$Q_{St} = I_G \cdot \tau \tag{16}$$

must be stored within the lowly doped region at this operating condition. Even at a relatively low carrier lifetime $\tau = 1$ µs this represents a carrier concentration which is about two orders of magnitude higher than the doping concentration. Under these circumstances the source-to-drain current increases strongly even at low drain voltages and starts to compensate the gate-to-drain current. At about $V_D = 10$ mV, a point is reached where, due to this compensation, the drain current reverses its direction, although the gate-to-drain current component is only slightly decreased. At even higher drain voltages the source-to-drain current is further increased, but now the reduction of forward bias of the gate-to-drain diode becomes noticeable in a considerable reduction of this diode current. The amount of stored charges decreases correspondingly and leads to an increase in the channel resistance. The drain current, therefore, stays approximately constant with increasing V_D in the range from 0.1 V to some volts. Only the constant gate-to-source diode current flows through the gate electrode in this V_D range. The operating conditions resemble those of a BPT, and Fig. 17 demonstrates that at a constant voltage $V_D = 0.5$ V a comparable current amplification is obtained.

The BSIT prepared by Y. Mochida and T. Matsujama[12] is used to exemplify the necessary process steps as follows:

(1) Epitaxial growth of a layer 2.6 µm thick with a donor concentration of 4.5 $\cdot$ 10^{13} cm^{-3} on a n^+-substrate of 0.01 Ω cm resistivity.

(2) Boron diffusion using a mask of silicon nitride with windows 3 µm in width. Oxidation using the same mask. During this process the oxide spreads, lifting off the silicon nitride. The width of the part covered by silicon nitride was varied from 4 µm in steps of 1 µm up to 11 µm.

(3) After removal of the silicon nitride film, the surface was covered by an n^+ doped polysilicon layer as a diffusion source, as shown in Fig. 15. The length of a single source region amounts to a little less than 260 µm and there were a total of 79 source lines on a chip. The total source length was 2.05 cm and the chip size amounted to 800 x 520 µm^2.

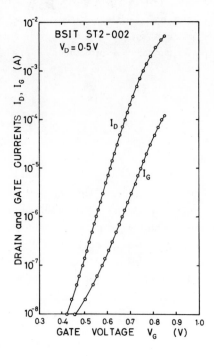

Fig. 17   Semilog plot of drain and gate current vs. gate voltage.

When the width of the silicon nitride was broader than 9 μm, the SIT produced was not pinched off. For V_G = 0 (gate short circuited to source) the drain current starts from an ohmic dependence on V_D. However, when the silicon nitride mask was narrower than 8 μm, the SIT was pinched off even with the gate short circuited to the source, and the I-V characteristics exhibited an exponential dependence.

Oscilloscope traces of the switching waveforms of a BSIT with a silicon nitride mask of 5 μm are shown in Fig. 18. A drain voltage of 21 V was applied via a 1 Ω load resistance and the gate continously biased to 0 V, -3.0 V, and -5.0 V, as shown in Figs. 18a, 18b, and 18c respectively. During a pulse of 1.2 μs duration the gate was forward biased to obtain I_D = 20 A. At the end of the pulse the current was switched off after delay times of 220 ns, 90 ns and 50 ns, respectively. The turn-on time of the BSIT was quite fast while the turn-off speed was limited by the charge-storage effect. The storage effect becomes negligibly small if the forward gate bias amounts to less than about 0.8 V. A forward voltage drop of 0.6 V was obtained at 20 A, which correspond to a current density of 2.5 - 3 x 10^4 A cm^{-2}. At a pulse current of 30 A, the voltage drop increased to 0.8 V and the turn-off delay times amounted to 440 ns, 170 ns and 110 ns for V_G = 0, - 3.0 V and - 6.0 V, respectively. The BSIT represents an excellent linear amplifying device, with low on-resistance which seems to be very useful in high-current, high-speed applications.

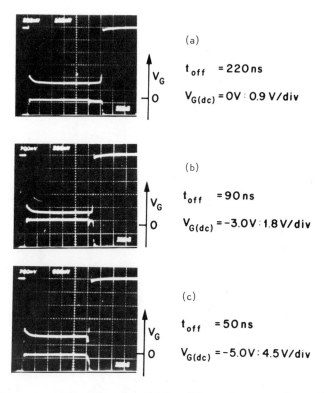

(a)

t_{off} = 220 ns

$V_{G(dc)}$ = 0V : 0.9 V/div

(b)

t_{off} = 90 ns

$V_{G(dc)}$ = -3.0V : 1.8 V/div

(c)

t_{off} = 50 ns

$V_{G(dc)}$ = -5.0V : 4.5 V/div

Fig. 18   Switching waveforms for the BSIT. (Upper trace: voltage (4.55 V/div); lower trace: gate voltage V_G. Time scale: 200 ns/div. R_L = 1 Ω, V_s = 21 V.

5. STATIC-INDUCTION THYRISTOR

Thyristors are well suited to high-power applications, since they combine a high blocking capability and a low on-resistance. The drawback of these components is that they
cannot switch off a d.c. current. Complete switching capability can be achieved by
implementing the operation principle to control a barrier height by electrostatic induction into a thyristor structure. To achieve this goal, the p-base layer is replaced by
an array of highly doped gate areas. Three different types of SI thyristors are
shown in Fig. 19. Part a) is a surface-gate type, which is the most easily produced
and allows the distributed gate resistance to be decreased by surface metallization.
Part b) shows a step structure and part c) a buried-gate type, which is convenient
for preventing surface breakdown. Both b) and c) allow very narrow channel width
to be achieved.

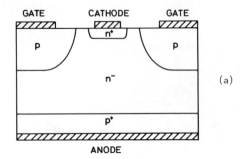

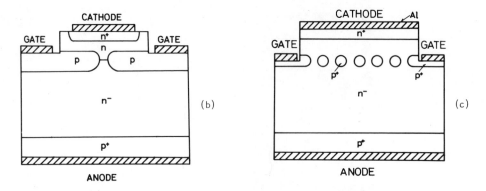

Fig. 19   Cross-sectional view of three types of SIT (a) surface-gate type, (b) step
structure, and (c) buried-gate type.

As in normal SIT's the barrier height in the channel at the center between neighboring gates depends upon the channel width and the gate-to-cathode voltage. With a
short-circuited gate a distinction is made between normally-off and normally-on thyris-

tors. When the spacing is small enough, so that the channel is pinched off due to the diffusion potential, then no electrons are injected into the lowly doped n-base region. Even if holes should flow back from the anode side, they would be absorbed at the gate and could not lead to an injection of electrons at the cathode. According to equation (12) the thyristor will stay in the blocking state, as long as $V_{G'} = \eta \ (V_G + \frac{1}{\mu} V_A)$ is kept negative. This condition requires a large voltage-amplification factor or a high negative gate bias which, however, is limited by the gate-to-source breakdown voltage.

With application of a positive gate voltage, the barrier is reduced and finally vanishes completely and the thyristor switches on in the same way as the corresponding forward-biased diode. The switch-on transient can even be improved compared to a diode by injection of holes from the gate area. The same turn-on process occurs in a normally-on Si-thyristor upon application of a positive anode voltage at zero gate bias. In such a device the channel width is wider than the extent of the depletion layers surrounding both gates due to the diffusion potential. In this case a negative gate voltage is required to pinch off the channel and thereby control a stable blocking state at positive anode voltage. In the on-state, the n-base is completely swamped by injected carriers. With an open contact the gate exhibits a floating potential between that of the cathode and the anode. If the gate potential is kept positive, it is generally the case that an extra gate current will flow. A reversal to negative V_G results in an extraction of holes from the stored charges and the feedback from anode- and cathode injection is interrupted. The potential barrier builds up again, and the device returns to the blocking state.

The above description explains the importance of the series resistance of the distributed gate structure for the switching behavior. In the case of switch-on, either the capacitance of the p-n junction has to be charged up or even holes injected into the surrounding n-region in a very short time. For this reason a gate-current pulse is needed. With a high series gate resistance a considerable voltage drop occurs, requiring a strong gate drive and causing inhomogeneous switching along the gate lines and additional turn-on losses. The situation is even worse in the case of switch-off, since the n-base is charged up with carriers by several orders of magnitude compared to the doping concentration. Usually a gate drive of only a few tens of volts is used, significantly below the gate-to-source breakdown voltage. Under these circumstances the gate resistance will often limit the negative gate current during turn-off. Only a gradual decrease of carrier concentration is possible resulting in correspondingly long turn-off times.

It is only in the on-state that benefit can be derived from the gate series resistance. In this case it limits the gate current, when the gate drive is kept at a constant potential, and this improves the efficiency. The optimum performance is obtained when the internal gate resistance is as small as possible whilst the gate electrode is connected to the drive via a parallel circuit of a resistor and a capacitor.

To reduce the turn-off time, Terasawa et al.[13] introduced an SI-thyristor with a shorted p^+-emitter structure as shown in Fig. 20a. The injection of holes from the periphery of the anode is avoided, and a reduction of turn-off time and leakage current at high temperatures is achieved. The 2.5-kV device, shown in Fig. 20b, was constructed. The chip has a diameter of 30 mm and the average current rating is

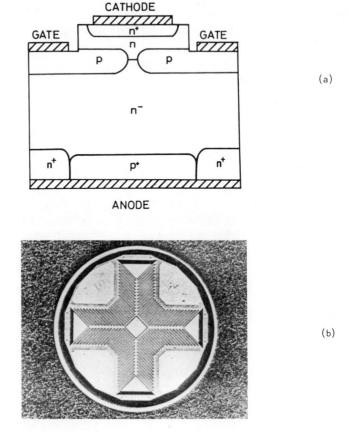

Fig. 20   (a) Cross-sectional view of SIT, (b) top view of 2.5-kV, 100-A device. (30 mm Ø).

100 A. The device has 120 cathode fingers with a total area of about 0.5 cm^2. The switching characteristics are shown in Fig. 21a. 100 A anode current can be switched off within 6 μs when -30 V are applied to the gate. The peak gate current was about 16 A. Thus, even measured at the pulse peak, a turn-off current gain of 6 and a voltage-amplification factor of 30 are obtained. Fig. 21b represents the turn-off time and the transient-current gain versus anode current. As a rough approximation, the total storage charge is proportional to the anode current and, therefore, both the

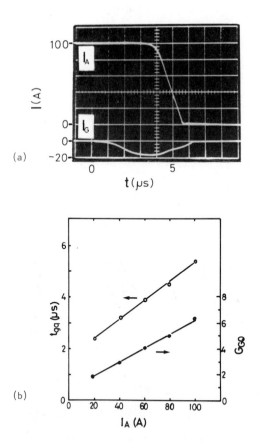

Fig. 21   (a) Switching waveform of the anode gate currents during turn-off of 2.5-kV, 100-A device. (b) the turn-off time and turn-off current gain vs. anode current.[16]

turn-off time and the transient-current gain vary linearly with anode current. The dV/dt-capability was about 1500 V/μs, and dI/dt values were determined to be about 10 times larger than those for conventional devices at a temperature of 125°C. With the gate short-circuited to the source, the blocking voltage was to 500 V. Applying a gate voltage of -10 V, an anode voltage of 3.0 kV could be blocked at a temperature of 25°C, and an anode voltage of 2.5 kV at a leakage current of about 4 mA at a junction temperature of 150°C.

By way of example, the fabrication technology for the SI-thyristor shown in Fig. 22 is described:

(1) An n-type silicon substrate with (111) orientation was used.

(2) At the back surface the anode emitter heavily doped with boron was diffused, followed by phosphorus diffusion of the short regions.

(3) The gate structure was produced by boron diffusion through an oxide mask on the front surface.

(4) After removal of the oxide mask, an epitaxial layer of n-doped silicon was deposited with an increase of the donor concentration at the end of the process.

(5) Phosphorus was selectively diffused before the evaporation of the cathode electrode.

(6) Part of the epitaxial layer was etched away to expose the contact region of the gate electrode.

Fig. 22   Top view at 600-V. 20-A device (6.3 x 6.3 mm^2) of Terasawa et al.[16]

The chip shown in Fig. 22a has a size of 6.3 x 6.3 mm^2 and 74 cathode fingers with a total area of about 0.06 cm^2. The device is capable of carrying an average current of 20 A, and the blocking voltage is 800 V with V_G = -30 V applied to the gate electrode. The minimum holding current is about 2 A and the turn-off time about 10 μs.

The turn-off time of SI-thyristors can be greatly reduced by introducing a second gate array close to the anode emitter.[6] As sketched previously, the stored carriers are extracted during the turn-off time. But even after the interruption of the injection the cathode emitter, the injection at the anode continues. The introduction of a second gate structure permits a similar sweep-out operation near the anode emitter. The second gate is coupled via a transformer to the pulse generator and, with the application of a small voltage pulse, turn-off times of less than 1 μs are achievable.

6. CONCLUSION

Experimental investigations of SITs, BSITs and SI-thyristors have revealed very promising electrical characteristics. The SITs seem to be especially well suited to high-frequency applications at low- to moderate-power levels. BSITs and SI-thyristors have advantages if low on-resistance is required, e.g. in the higher power range. They do not permit such high-frequency application as SIT's, but seem superior to conventional devices.

REFERENCES

1. J.E. Lilienfeld: U.S. Patent No. 1745175 (1930), No. 1900018 (1933).

2. O. Heil: British Patent No. 439457 (Issued December 1935).

3. W. Shockley, "A Unipolar Field-Effect Transistor," Proc. IRE, 40 (November 1952) 1365-1376.

4. G.C. Dacey and I.M. Ross, "Unipolar Field-Effect Transistor," Proc. IRE, 41 (August 1953) 970-979.

   G.C. Dacey and I.M. Ross, "The Field-Effect Transistor," BST J, 34 (1955) 1149-1189.

5. J. Nishizawa and Y. Watanabe: Japanese Patent 205068 Published Number 28 - 6077 (Fig. 15), Application date December 1950.

6. J. Nishizawa, T. Terasaki, and J. Shibata, "Field-Effect Transistor Versus Analog Transistor (Static Induction Transistor)," IEEE Trans. Electron Devices, ED-22, No. 4 (April 1975) 185-197.

7. J. Nishizawa, T. Terasaki, and J. Shibata, "Field-Effect Transistor Versus Analot Transistor (Static Induction Transistor)," IEEE Trans. Electron Devices, ED-22, Not. 4 (April 1975) 185-197.

J. Nishizawa, "SIT Integrated Circuits," 1977, Proc. of SRI Conference on Semiconductors, Chapter 6, Handotai Kenkyu (Semiconductor Electronics) Vol. 15, Chap. 6 (Kogyo-Chosakai, 1978).

J. Nishizawa and K. Yamamoto, "High-Frequency High-Power Static Induction Transistor," IEEE Trans. Electron Devices, ED-25 (March 1978) 314-322.

J. Nishizawa, "Recent Progress and Potential of SIT (Invited)," Proc. of the 11th Conference (1979 International) on Solid State Devices, Tokyo, 1979; Japanese Journal of Applied Physics, 19 (1980), Supplement 19-1, 3-11.

8. M.S. Shur and L.F. Eastman, "Ballistic Transport in Semiconductor at Low Temperature for Low-Power High-Speed Logic," IEEE Trans. Electron Devices, ED-26, No. 11 (November 1979) 1677-1683.

9. C.Y. Chen, A.Y. Cho, P. Garbinski, and G. Bethea, "A Majority-Carrier Photodetector Grown by Molecular Beam Epitaxy," 5th International Conference on Vapor Growth and Epitaxy (ICVGE-5), Abstracts (July 1981) 62-63.

10. J. Nishizawa, "High Frequency Base Resistance, Emitter Cut Off and Maximum Power in the Junction Type Transistor," Trans. IECE of Japan, 44, No. 5 (May 1961) 767-776.

11. Y. Kajiwra, Y. Higaki, M. Kato, and S. Mitsui, "Analysis of Operation Mechanism of a Static Induction Transistor Using a Cross-Type Model," Trans. IECE of Japan, J63-C, No. 8 (August 1980) 529-536.

12. J. Nishizawa, T. Ohmi, Y. Mochida, T. Matsuyama, and S. Iida, "Bipolar Mode Static Induction Transistor," International Electron Device Meeting, Technical Digest (December 1978) 676-679.

13. Y. Terasawa, M. Miyata, S. Murakami, T. Nagano, and M. Okamura, "High Power Static Induction Thyristor," International Electron Device Meeting, Technical Digest (December 1979) 250-253.

Y. Terasawa, "Semiconductor Technologies," Japan Annual Review in Electronics, Computors and Telecommunication, in press.

DISCUSSION

(Chairman: Prof. P. Leturcq, Institut Nationale des Sciences Appliquées, Toulouse)

A. Jaecklin (BBC Brown, Boveri & Co. Ltd., Baden)

The devices you have shown, especially Hitachi's high-voltage device, are very close to the specifications on the gate turn-off thyristors we discussed yesterday. Can you compare the two?

J. Nishizawa (Tohoku University, Sendai)

To achieve high-speed operation of a thyristor one has to reduce the carrier life-time in the device. The switching time is then controlled by the extraction of carriers by the gate and the device can be produced with a long carrier lifetime. It can be simultaneously rather fast and have low dissipation.

A. Jaecklin

How about practical applications?

J. Nishizawa

My feeling is that this device can be applied to d.c. power transmission in the future. The speed is already sufficient to realize induction heaters. With these thyristors we can chop at frequencies up to 450 kHz and the efficiency seems to be greater than 90% even in this case.

M. Adler (General Electric, Schenectady)

Is it correct that the 2.5 kV, 500 A thyristor turns off in 5 µs with a gate current of only 5 A?

J. Nishizawa

Yes, this has been published by Hitachi and some details are given in my paper.

M. Adler

Some results were published for an earlier device exhibiting a turn-off gain of about 4 or 5 instead of 100.

J. Nishizawa

That was not so good and has been improved upon since then.

H. Becke (Bell Laboratories, Murray Hill)

I do not understand the phenomenon of the turn-off gain of the static induction thyristor. From a simple physical point of view, since all the carriers go out the gate, the gate current for turn off over the total current flowing must be just the ratio of electron to hole mobility and I cannot understand how you can achieve a current gain of 100.

J. Nishizawa

Most important are the gate structure and the thickness of the base. The potential just stops the injection from one side. For these reasons I believe we can get such high current gain.

BIPOLAR TRANSISTORS

P. L. HOWER[*]
Unitrode Corp. Watertown, Mass. USA

SUMMARY

Despite its age, the bipolar transistor is still one of the major switching components used for power electronics applications. For those cases where controlled turn-off is required, it is probably the dominant component. In recent years this dominance has been challenged by the emergence of power MOSFETs and GTOs as possible alternatives for power-switching applications. In this paper the three devices are compared. The bipolar receives the most attention, with particular emphasis on performance limitations imposed by the physical properties of the starting material (silicon) and processing techniques. The failure modes and turn-off behavior are also considered. The paper concludes with a discussion of possible future trends in the design and fabrication of power bipolar transistors.

1. INTRODUCTION

Any investigation into the history of bipolar power transistors quickly reveals that present-day devices bear only a remote resemblance to the original point-contact and junction transistors as far as mechanical form is concerned. The motivation of the early workers was to develop devices for communications applications, that is, small-signal amplification was of primary interest. Technological efforts were directed towards increasing the operating frequency and reducing noise. Despite these efforts in the linear amplification regime, it was quickly noted that the early devices were vastly superior to the vacuum-tube switches then being used in digital computers. Technological advances came quickly, and within a few years computer applications dominated the output of bipolar production.

1.1 The Slow Start of Power Electronics

Despite the early recognition of the switching capabilities of bipolars, it was almost two decades before the device was routinely used to perform power-switching functions. The reasons for this delay are not clear cut. One can list a number of possibilities having to do with both technological and economic factors. For example, early

---

* previously with Westinghouse R&D Center, Pittsburgh, Pennsylvania, USA

devices were somewhat limited in blocking voltage and were rather small in area, and in consequence their "V-I switching products" were small. Transistors were limited to applications with output powers of a few watts or less.

Also, devices tended to fail catastrophically, particularly when called upon to interrupt inductive currents. Early power transistors were costly compared to competitive methods of switching power, such as relays and vacuum tubes, and device failure tended to discourage their use in power-switching applications.

Another inhibiting factor was that power electronics had yet to become the widely developed art which exists today. Power conversion was a specialty practised by a relatively small number of workers. During this period computers were becoming larger, but not until the late 1960's had they reached the size where thermal considerations justified the use of switching regulators in place of conventional series regulators. By increasing the overall efficiency, switching regulators greatly reduced the input power requirements and thereby eased the thermal-management problem.

By the early 1970's considerable progress had been made in solving many power-conversion problems. Therefore, when the need for efficient energy management was suddenly brought home by the oil embargo and other related crises of the 1970's the previously arcane art of power-conversion techniques began to receive major attention.

Energy-saving techniques that were too expensive to implement using energy costs of the 1960's suddenly became attractive under the new energy rules that began to emerge in the 1970's. As a result of these and related changes, demands for improvements in the design and characterization of switching components received new emphasis. Since there are still a number of major problems to be solved in the design and fabrication of power-switching devices, it is likely that these demands will continue for some time into the future.

1.2 Power Transistor Milestones

As one might gather from the list of problem areas already given, the path of bipolar power-transistor progress is not straightforward. Some of the significant milestones are listed in Fig. 1. Following the discovery of the point-contact transistor, the development of practical junction transistors was a major accomplishment and provided the foundation from which subsequent power-device development proceeded.

Early power transistors had a lightly doped base with alloyed regrowth regions forming the collector- and emitter-doped layers. Both germanium and silicon devices were made this way. It was quickly realized that silicon was the proper choice for blocking voltages above about 75 V.

Power Transistor Milestones

Bardeen, Brattain, and Shockley	1948	Point-contact transistor
Shockley	c. 1949	Junction transistors
Sparks	1950	Grown-junction transistor
Saby; Law, Mueller, Pankove, and Armstrong	c. 1951	Alloy-junction transistor
Kromer	1953	Drift transistor
Early	1954	p-n-i-p,n-p-i-p transistor
Christiensen and Teal;	1954	Si-epitaxial transistor
Theurer et al.	1960	Si-epitaxial transistor
Lee, Tannenbaum, and Thomas	1956	Diffused-base transistor
Texas Instruments	c. 1956	Grown-junction Si transistor
Thorton and Simmons	1958	Second breakdown
Fairchild	c. 1958	Planar transistor
RCA	c. 1959	Single-diffused transistor
Lucas Electric; Delco	1960	Triple-diffused transistor
Scarlett and Shockley	1963	Hot spots in transistors
Grutchfield and Moutoux	1966	Current-mode second breakdown
Westinghouse	c. 1968	Adaptation of high-power device technology to transistors
Hower and Reddi	1970	Avalanche injection as trigger for second breakdown
Denning and Moe	1970	Pi-Nu epitaxial transistor

Fig. 1   List of important events in the development of power transistors.

Diffusion of impurities in transistors was originally used to achieve a thin metallurgical base. Such transistors were aimed at communications applications. Diffusion was turned to advantage in the development of the single-diffused power transistor which has the profile shown in Fig. 2. This profile resembles that of the alloy transistor, and from this standpoint not much was changed. However, there were some characteristics of impurity diffusion that made it a very desirable innovation. For example, diffusion allowed finer control of the basewidth than was possible with most alloying techniques. Perhaps even more important than the degree of control was the fact that when impurity-diffusion methods were combined with the photolithographic masking and metallization techniques under development in the late 1950's it became possible to achieve the amount of emitter-base interdigitation required for high-current power transistors.

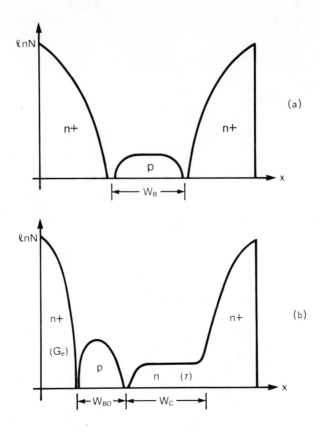

Fig. 2   Impurity profiles for (a) single-diffused and (b) triple-diffused profile.

The original proposal of a lightly doped collector region by Early (1954) was aimed at achieving a high-frequency, small-signal transistor. With the development of epitaxial-growth methods, Early's idea became easier to implement. This development was important because lightly doped collector transistors have an advantage over single-diffused transistors from the standpoint of the breakdown-voltage tradeoff that pertains to both devices. This tradeoff will be discussed in more detail later. At this point it is only necessary to note that as the breakdown voltage is increased, various penalties must be paid in terms of decreased collector current, decreased gain, or increased device size.

Impurity-diffusion methods were also used to achieve a lightly doped collector profile (see Fig. 2(b)). So-called triple-diffused transistors were developed in the mid and early 1960's for electronic ignition in automobiles and deflection circuits in TV receivers. The ignition transistors are of particular interest because their breakdown voltage, which was in the 300 to 500 V range, made them suitable for use in switching regulators which operate directly off the line. These transistors were readily available at moderate cost and aided significantly in the development of early switching-regulator designs.

During the development of these new high-voltage transistors, there were also a number of advances made in the understanding of device behavior. It was shown by Scarlett and Shockley[1] that transistors could become thermally unstable forming "hot spots" that were a prelude to a mysterious failure mechanism dubbed "second breakdown". Device designers and users learned to deal with this problem by developing a "Forward Safe Operating Area" which defines regions in the V_{CE}, I_C plane that are free of second breakdown for a particular pulse time.

Failure during inductive turn-off was a more difficult problem to deal with, and it is now generally agreed that the triggering mechanism in most applications is avalanche injection[2] which is nonthermal in origin. Device users circumvented some of the limitations of transistors in this mode by using protective circuits which either clamped V_{CE} or limited its rate of increase during turn-off.

During the 1960's it was noticed that high-voltage transistors exhibited behavior that deviated in a number of ways from the "classical" single-diffused or alloy transistors. One difference was that the common-emitter current gain h_{FE} changed as a function of both V_{CE} and I_C in high-voltage transistors. On the collector characteristic this behavior is displayed by the constant curves becoming increasingly compressed as V_{CE} approaches the origin. In a lightly doped base transistor, the constant I_B lines remain nearly horizontal as shown in Fig. 3.

The physical reason for this "quasi-saturation" behavior was finally identified as being due to a decrease in emitter injection efficiency which accompanies the "base-widening" process.[3,4] Because the width of the "current-induced base" depends on V_{CE} and I_C, the cut-off frequency f_T was observed to be dependent on bias in a way that was also quite uncharacteristic of a classical, lightly doped base transistor. Similarly, switching times were observed to increase rather dramatically when the switching locus extended beyond the active region into the quasi-saturation region.

With time, many of the puzzling aspects of the high-voltage transistors were resolved. Despite this progress, a number of questions remain unanswered. In particular, behavior during turn-off is only partially understood. Some of these questions will be considered later in the paper.

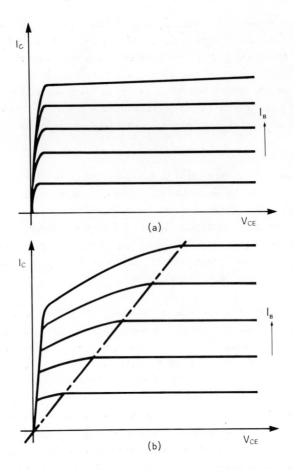

Fig. 3   Collector characteristics at low voltages for (a) single-diffused and (b) triple-diffused transistor.

During the early 1970's, transistor users made major advances in switching-regulator and other power-conversion systems. It was realized that failure during turn-off could be avoided by tailoring the switching locus by various circuit techniques.

Later in the 1970's, drastically increased energy costs emphasized the need for improved devices and power-conversion methods. It was generally recognized, however, that the main switching elements as well as the control logic would most likely consist of silicon devices.

One improvement took the form of increasing transistor current-handling capability by scaling up the conducting area using high-power thyristor-fabrication and assembly techniques. This effort resulted in devices capable of controlling in excess of 10 kVA of switching power and permitted switching-regulator techniques to be applied to higher power applications. An example of such a "large-area" transistor will be described later in the paper.

2. CANDIDATE SWITCHES FOR CONTROLLED TURN-OFF

At present there are three readily available devices for performing the switching function. These are the power MOSFET, the gate turn-off thyristor (GTO), and the bipolar transistor. Although we could also add conventional thyristors when operated in a force-commutated turn-off mode, the size and cost of the accompanying circuit elements limit this type of switch to special situations. Cross sections of the three devices are shown in Fig. 4, with the shaded regions indicating the approximate location of excess charge when the device is in the "on" state. At this point it is worth noting some of the important differences between the three devices. Table 1 lists various properties which are of interest for power-switching devices.

TABLE 1

Comparison of Silicon Power Switches

Property	MOSFET	Bipolar	GTO
Relative amount of charge stored in on-state (C/cm^3)	small	moderate	large
Practical range of on-state current density (450-V design) (A/cm^2)	50/200	50/200	500/800
On-state voltage drop (V)	3 to 6	0.3 to 1	2 to 3
Storage time (μs)	< 0.2	1 to 10	1 to 10
Reverse safe operating area	excellent	limited	limited
Drive power/complexity	small	moderate	large
Conduction losses (W/cm^2)	large	small	small
Switching loss per pulse (mJ/cm^2)	small	moderate	moderate to large
Active area/die area	~ 0.5	0.5	0.3
Maximum junction temperature (°C)	150	175 to 190	125
Process technology	integrated circuit	conventional	conventional

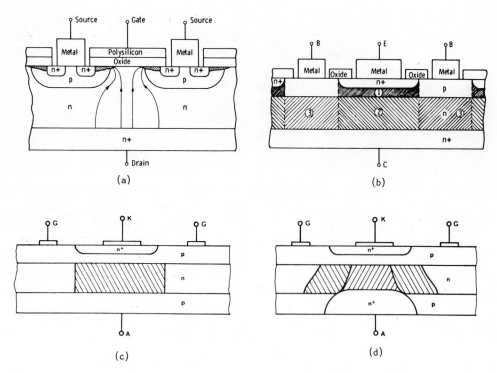

Fig. 4   Cross sections of the working areas of various candidate switches: (a) MOS-
FET("double-diffused"), (b) bipolar, (c) and (d) GTO.

Of the three devices the MOSFET, which makes use of integrated circuit technology,
is the most recent. Although the bipolar and GTO use more conventional technology,
a recent bipolar design has been developed using IC layout methods.[5] Two cross sections are shown for the GTO: the first is a conventional design and is used when
blocking in the reverse direction is required;[6] the second GTO is a design due to
Becke and Neilson;[7] and is essentially a parallel combination of a bipolar transistor
and a thyristor. The latter design minimizes some of the plasma localization problems
that can occur during turn-off of a conventional GTO.

All devices require a certain thickness and doping of the n-layer to support the required blocking voltage, but differ markedly in the amount and location of excess
charge that is supplied to achieve the "on" or conducting state. In the bipolar this
charge modulates the conductivity of the collector n-layer thereby achieving a reduction in on-state voltage drop over the MOSFET. In the GTO additional charge is
stored due to regenerative transistor action similar to that of a thyristor and in
consequence large "on" current densities can be achieved in the GTO at moderate
forward voltages. The penalty paid is that a considerable amount of reverse-drive

current must be supplied to achieve the desired turn-off time. Also, localizations of current density can occur during turn-off that lead to device failure.

These undesirable charge localizations also apply to the bipolar but to a lesser extent. Using a somewhat simplistic viewpoint, one can say that there is less charge to "man-

age" during turn-off of a bipolar. For the MOSFET, even less charge is involved and turn-off can be achieved without much difficulty. For the MOSFET, the major design precaution is the need to suppress the transistor action of the parasitic bipolar.

One way to compare the three devices from a simplified viewpoint is shown in Fig. 5.

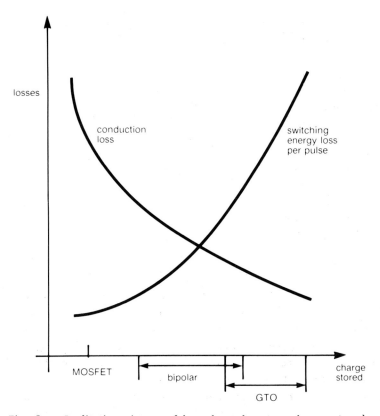

Fig. 5   Qualitative picture of loss dependence on charge stored.

It can be seen that the bipolar lies between the MOSFET and GTO in terms of all three quantities: conduction loss, switching energy loss, and charge stored. The degree of drive-circuit complexity also increases with the amount of charge stored.

While helpful for describing the devices in general terms, this simplistic picture is limited when comparing them on a more quantitative basis. Some of the results of a more detailed comparison[8] between the MOSFET and bipolar are shown in Figs. 6 and 7. From these figures it can be seen that the bipolar will always have lower conduction losses than the MOSFET but, because of its increased switching losses, the bipolar total-loss curve will always cross the MOSFET curve as frequency increases. For present devices, the crossover frequency is approximately 15 kHz and is relatively insensitive to junction temperature.

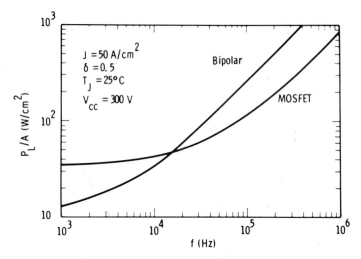

Fig. 6   Power-loss density vs. switching frequency for the bipolar and MOSFET $(T_J = 25°C)$.

In the future we can expect technological improvements in both devices. It is unlikely that the conduction loss of the MOSFET can be reduced much below the value shown in Figs. 6 and 7 because the quantity shown here is due only to the n-layer resistance which is necessary for a given BV_{DS}. Improvements in cell layout and reduced channel resistance will only allow a closer approach to this value. Therefore the left-hand portion of the MOSFET curve will remain very nearly fixed. Although a similar statement could be made about the bipolar, here there is some hope for reduction in the on-losses, due to improved RSOA designs and also improved emitters. This topic is considered in more detail in the next section.

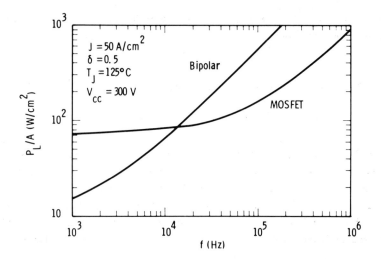

Fig. 7   Power-loss density vs. switching frequency for the bipolar and MOSFET
          (T_J = 125°C).

For both devices, the right-hand portions of Figs. 6 and 7 can be expected to shift to
smaller loss values with improvements in technology. Therefore a possible set of future
curves could be those indicated by the broken lines of Fig. 8. As indicated in this

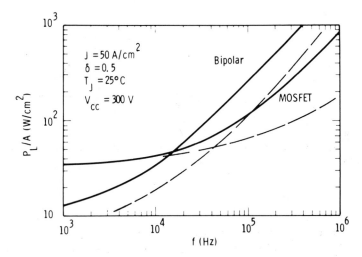

Fig. 8   Comparison with curves of Fig. 6: broken lines indicate future possibilities.

figure, the crossover frequency between the bipolar and MOSFET can only increase with technological advances.

3. BIPOLAR CHARACTERISTICS AND FURTHER COMPARISONS

3.1 G vs. V_{CEO}(sus) Tradeoff

One of the fundamental properties of a bipolar transistor is that the common-emitter current gain h_{FE} becomes inversely proportional to collector current as I_C is increased. That is the product $h_{FE}I_C$ approaches a constant, which we define as G. For example, if $h_{FE} = 5$ at $I_C = 100$ A, then G = 500 A, which is the case illustrated in Fig. 9. In this figure, two transistor designs are shown both having G = 500 A but with different peak-gain values (h_{FEO}) due to differences in the base impurity profile.

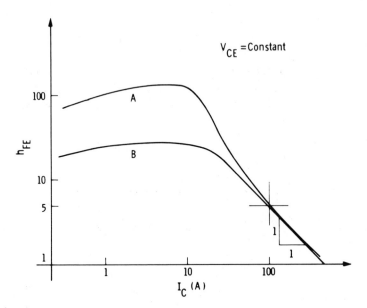

Fig. 9   Bipolar common-emitter current gain vs. collector current for two different base profiles.

For most switching applications, the circuit designer is interested in the value of G rather than h_{FEO}. This means that the transistor designer may adjust h_{FEO} (within limits) to improve other characteristics, such as fall time, storage time, or sustaining voltage. For most transistor designs G can be approximated by[9]

$$G = h_{FE} \, I_C = \frac{4q \, A_E \, G_e \, D_C^2}{W_C^2}$$

(1)

In this equation A_E is the effective emitter area, W_C the collector width (see Fig. 2(b)), D_C the high-level diffusion coefficient, and G_e the "emitter Gummel number" which characterizes the injection capabilities of the emitter.*

Some recent work has shown that G_e is linked to the recombination lifetime in the collector. Fig. 10 shows a plot of measured G_e versus open-circuit-decay lifetimes measured at the base-collector junction terminals. While there is considerable scatter in the data, it can be seen that G_e increases towards some asymptotic value with increasing lifetime.

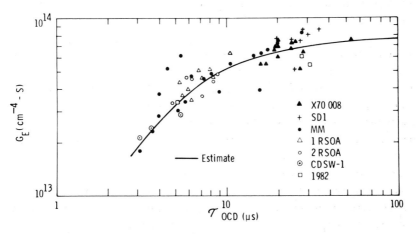

Fig. 10   Measured emitter Gummel number vs. open-circuit-decay lifetime for various transistors.

*
Frequently the recombination parameters $h(n^+)$ and $h(p^+)$ are used to characterize the back-injection properties of heavily doped layers.[10] In this context, $G_e = 1/h(n^+)$.

Other work by Oh-uchi et al.[11] has shown that it is possible to make rather dramatic increases in G_e by making shallow emitter structures which are contacted by various polysilicon and oxygen-doped polysilicon layers which have the effect of acting like a heterojunction emitter. If such an emitter could be incorporated into a conventional power-bipolar structure, major increases in the gain-current product would be achieved. For the purposes of this paper, however, we shall use G_e values which are more in line with Fig. 10 for estimates of bipolar performance.

For a given process and mask design, G_e and A_E will be fixed. Equation (1) then suggests that W_C should be reduced to make G as large as possible. Unfortunately, decreasing W_C decreases the open-base sustaining voltage, $V_{CEO}(sus)$. In an optimum design

$$V_{CEO}(sus) = K_1 W_C \tag{2}$$

where $K_1 \cong 10$ V/µm for a typical design. Thus increasing $V_{CEO}(sus)$ requires larger

W_C and according to equation (1), G must decrease. For example, doubling the voltage decreases G to a quarter of its previous value. A more precise analysis[12] shows that G is actually inversely proportional to $V_{CEO}(sus)^{2.3}$ which is a more severe tradeoff than is predicted by the approximations of equations (1) and (2). That is, doubling the voltage gives a fivefold reduction of G.

Calculated G vs. $V_{CEO}(sus)$ curves are plotted in Fig. 11 for different values of emitter area. TO-3 transistors are limited to A_E values of about 0.15 cm^2 or less. Larger area transistors, with A_E in excess of 1 cm^2, have been made using the packaging technology of high-power thyristors. These transistors are commercially available and are finding use where switching products in excess of 10 kVA are desired. The dot shown in Fig. 11 applies to a device that will be described later in the paper.

In connection with Fig. 11 it should be noted that A_E is the effective emitter area. The metallurgical emitter area A_{EM} must be larger than A_E by a ratio that is a function of the collector current and emitter stripe-width.[13] If A_E is to be within a desired fraction of A_{EM} the emitter width must be sufficiently small. This means that lower voltage designs must have increasingly finer geometries to achieve the G values of Fig. 11.

A second limitation is due to thermal dissipation requirements. Bipolar transistors are limited to typical on-state power densities in the range of 200 to 300 W/cm^2, for a 100 percent duty factor. This means that the larger G values of Fig. 11 may be achievable only on a pulse basis.

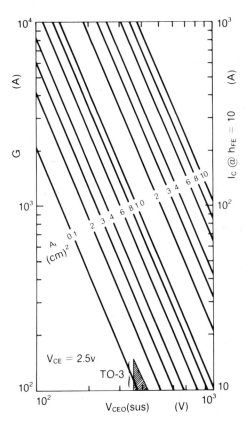

Fig. 11   Gain-current product vs. open-base sustaining voltage (parameter = emitter area).

3.2 Comparison with the MOSFET

Fig. 11 indicates that the tradeoff between blocking and on-state performance is rather severe for the bipolar. That is, for only a slight increase in V_{CEO}(sus), G suffers a large decrease. Although it might be thought that the MOSFET would be more favorable from this aspect, a similar type of calculation shows that the on-resistance of the MOSFET is proportional to $BV_{DS}^{2.3}$. This means that the MOSFET suffers from the same fundamental limits as the bipolar. These limits are functions of material properties such as mobility and ionization coefficients, and a plot of G and R_{on} would show the same proportionality with the blocking voltage V_{CEO}(sus) or BV_{DS}. At lower voltages both devices can run into a thermal limitation due to the necessity to remove heat in the on-state. In this case the MOSFET suffers a disadvantage due to the temperature dependence of the mobility. Temperature has only a minor influence on G for the bipolar.

To continue the comparison we calculate the current density that can be controlled by each device. For the plots of Figs. 12 and 13, both BV_{DS} and $V_{CEO}(sus)$ are set equal to 450 V. For the MOSFET there is a single resistive line while for the bipolar J_{on} depends upon the value of gain used. The resistive line at the left of the bipolar curves is due to the emitter metallization which is estimated to be 3 x 10^{-3} ohm-cm^2.

The MOSFET lines of Figs. 12 and 13 are unlikely to change, since they represent the best one can do for BV_{DS} = 450 V. That is, the channel resistance and junction FET action between adjacent "p-wells" has been ignored. For the bipolar curves, some

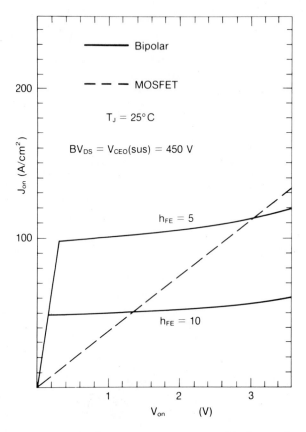

Fig. 12   On-state current density vs. voltage for a bipolar and a MOSFET (T_J = 25°C).

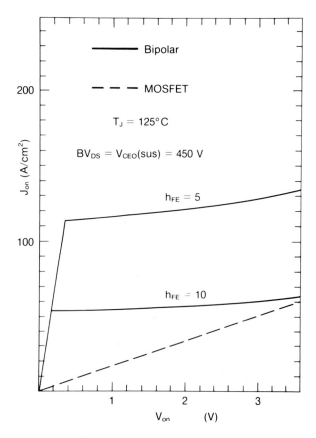

Fig. 13 On-state current density vs. voltage for a bipolar and a MOSFET (T_J = 125°C).

improvement can be expected because V_{CEO}(sus) can be exceeded in some cases. To demonstrate this possibility, the collector and drain characteristics of a MOSFET and a bipolar having the same n-layer thickness and doping are shown in Fig. 14. If the emitter can be prevented from injecting, the bipolar will be limited by its collector-base breakdown voltage BV_{CBO}, which in this case is equal to BV_{DS} of the MOSFET. As stated earlier, it is believed that improvements in device design and characterization will permit more efficient use of the region to the right of the V_{CEO}(sus) line. For the present examples, this means that the J_{on} curves for the bipolar will move upward.

Thus from the standpoint of the blocking voltage, on-state tradeoff, which determines the "d.c." performance, it is concluded that the bipolar will always have the advantage in terms of area of silicon required. For small devices, this advantage is of minor interest and other criteria such as drive power and circuit simplicity may cause the MOSFET to be the preferred device. As the power level increases, efficient use of die area becomes more important and the arguments proposed here begin to take effect.

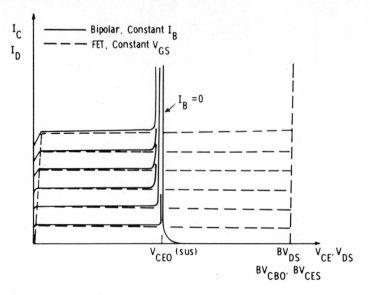

Fig. 14   Collector and drain characteristics for a bipolar and a MOSFET.

3.3 Maximum Temperature Limits

In the active region the bipolar can become thermally unstable due to positive feedback resulting from transistor action. This instability can occur at temperatures considerably less than the maximum junction temperature T_{Jmax} which is typically 175°C to 200°C for a bipolar. Transistor thermal instability leads to deviations from a constant power locus in the forward SOA diagram.

The MOSFET does not exhibit this type of thermal instability provided the parasitic bipolar has been properly suppressed. This means that the MOSFET should have a forward SOA determined by a T_{Jmax} in the range of 175°C to 200°C. At present this temperature range is rather optimistic since present MOSFETs are limited to T_{Jmax} values of approximately 150°C. This limit is imposed by the need for long-term stability of the gate oxide. Consequently, when comparing devices of the same size, the forward safe operating areas will appear approximately as shown in Fig. 15. It is assumed here that both devices have the same maximum current rating and the same thermal resistance.

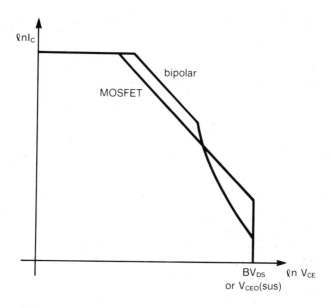

Fig. 15   Forward safe operating areas for a bipolar and a MOSFET having the same
thermal resistance.

Usually the forward SOA is of minor interest in switching applications since operation
in the active region is for relatively short time intervals. The limitation for switching
operation in both devices is imposed by the maximum temperature T_{Jmax}. This temperature, together with the device power losses and heat-sink capability, determine the
maximum frequency that can be switched. That is, there is a "safe operating frequency"[14] which increases as the power level switched decreases.

3.4 Single Bipolar vs. a Darlington

One objection to the bipolar is the need for a relatively large amount of base current
to keep the transistor in the on-state. Although current feedback schemes can be
used to circumvent this problem, the more traditional approach is to use a Darlington.

This configuration, which is shown in Fig. 16, has some disadvantages in that the
increase in current gain is achieved only during turn-on and V_{CE} in the on-state
can never be less than the V_{BE} of the output transistor. Also turn-off is a relatively

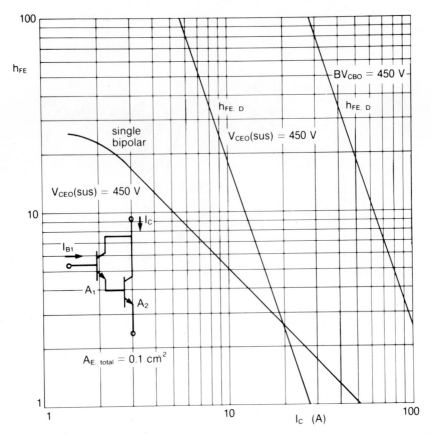

Fig. 16   Comparison of the h_{FE} fall-off behavior for a single bipolar and a Darlington
(blocking voltage = 450 V).

slow process unless reverse current is applied to the base of the output transistor by means of a bypass diode or a separate terminal.

For many applications, fast turn-off is not required. For example, certain motor-speed control applications require a switching frequency of only a few kilohertz. In this case the upper limit on voltage need not be $V_{CEO}(sus)$ but can be BV_{CBO} provided an appropriate snubber network is used. Therefore the Darlington has a considerably better blocking voltage, on-state tradeoff than does a single bipolar for these applications.

It is worth considering this problem in a little more detail. The gain of a single bipolar can be approximated by

$$h_{FE} = G/I_C \tag{3}$$

in the "fall-off" region, where G can be regarded as a figure-of-merit for a given $V_{CEO}(sus)$. The gain for a Darlington obeys a different approximation which can be shown to be

$$h_{FE,D} \cong G_1 \, G_2^{\,2}/I_C^{\,3} \tag{4}$$

where G_1 and G_2 are the gain-current products of the driver and output transistors. These terms are proportional to the emitter areas A_1 and A_2 of the driver and output transistors. From equation (4) it can be seen that the gain of a Darlington falls off with I_C much more rapidly than for a single bipolar.

A consideration of the problem of how to apportion the driver and output areas shows that the factor $G_1 G_2^{\,2}$ will be maximized when $A_2 = 2A_1$. In this case

$$h_{FE,D} = 0.15 \, (G/I_C)^3 \tag{5}$$

where use has bee made of the fact that

$$G = G_1 + G_2 \tag{6}$$

due to the requirement that A_1 and A_2 must add up to the area of the single bipolar being used for comparison. Fig. 16 shows a plot of h_{FE} vs. I_C for a single bipolar with $V_{CEO}(sus) = 450$ V and $A_E = 0.1$ cm^2 which gives G = 50 A. Also shown are gain curves for a Darlington of the same size with $V_{CEO}(sus) = 450$ V and $BV_{CBO} = 450$ V. Fig. 17 contains the same kind of comparison but for a blocking voltage of 900 V. From these curves one can see that the Darlington is the preferred device since it can control a considerably larger current at increased gain.

One drawback of the Darlington is that the output transistor is usually operated at a higher current density than would be normal for a single bipolar. As will be discussed in a subsequent section, this means that the device is more prone to failure during turn-off if the turn-off locus crosses into a region where avalanche generation is significant. A second drawback is that the Darlington cannot achieve the low voltage drop of the single bipolar. Therefore working at the large currents suggested by the curves of Figs. 16 and 17 may produce an undesirably large power loss in the on-state.

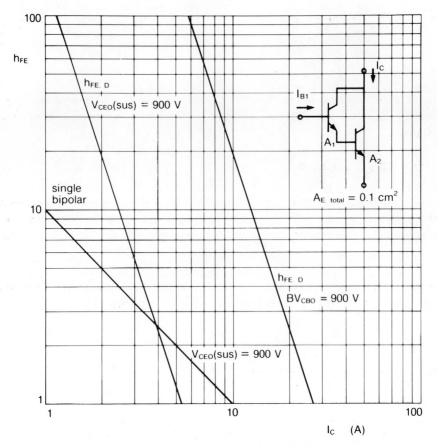

Fig. 17   Comparison of the h_{FE} fall-off behavior for a single bipolar and a Darlington (blocking voltage = 900 V).

Finally, even for the most optimistic case where the blocking voltage is due to BV_{CBO}, the Darlington still works at a lower current density than can be achieved with a GTO of similar voltage rating. For this reason GTO's are of continuing interest for lower frequency applications even though the drive circuits are somewhat more complicated than with a Darlington.

3.5 Large-Area Bipolar Transistors

In the past few years high-power thyristor packaging techniques have been adapted to the manufacture of large-area (> 1 cm^2) bipolar transistors. Recently, it has been shown that it is possible to make devices with metallurgical emitter areas of 3.2 cm^2 using these techniques.[15] The transistor element, which is 33 mm in diameter, is shown in Fig. 18. Measured h_{FE} vs. I_C is shown in Fig. 19 (a) and (b) for two devices with V_{CEO}(sus) = 428 V and 577 V, respectively. Typical turn-off waveforms for a clamped, inductive-load test circuit are shown in Fig. 20. This device is intended for use in a 20-kHz, 20-kVA inverter for aircraft applications.

Although the dominant market for bipolar power transistors is limited to the TO-3 and smaller sizes, the results presented here demonstrate that bipolars can be scaled up to handle power levels 20 to 30 times that available with a TO-3. For this reason, bipolar transistors present some interesting possibilities as switching components in future high-power applications.

Fig. 18   Photograph of a 33-mm dia. transistor element for packaging via high-power thyristor techniques.

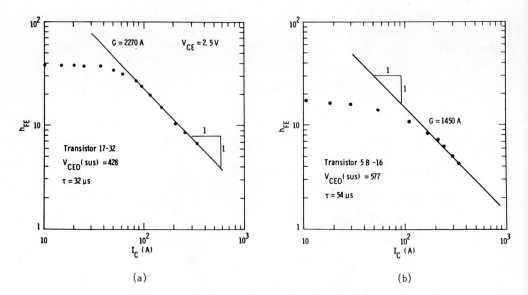

Fig. 19   Current gain vs. collector current for a transistor similar to Fig. 15.
          (a) V_{CEO}(sus) = 428 V, (b) V_{CEO}(sus) = 577 V.

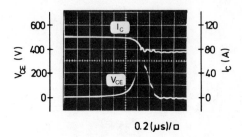

Fig. 20   Switching waveform for a clamped inductive circuit for a transistor similar
          to Fig. 18.

4. PROBLEM AREAS

4.1 Turn-Off Limitations

All three candidates, the MOSFET, bipolar, and GTO, suffer from the possibility of catastrophic failure during turn-off. In the MOSFET the problem can be minimized by reducing the gain of the parasitic bipolar and by designing in a well-distributed network of base-emitter shorts. Using these techniques the parasitic bipolar will not conduct during turn-off except for very large currents.

In both the GTO and bipolar, current density becomes localized during turn-off to the center of the emitter stripe.[16,17] In the bipolar, charge is initially removed from the region beneath the base contact which is labeled region 3 in Fig. 21. Following this

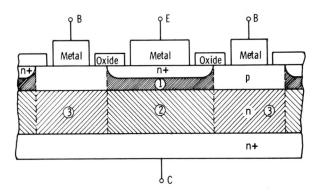

Fig. 21   Cross section showing excess charge stored in (1) remote base, (2) current-induced base, and (3) metallurgical base.

interval, charge is next removed from the "current-induced base", or region 2 of Fig. 21. During this interval, the current-induced base shrinks in both its lateral and vertical dimensions as indicated in Fig. 22. In the final stage of turn-off, the collector-base junction begins to block and excess charge is removed from the metallurgical base (region 1 of Fig. 21).

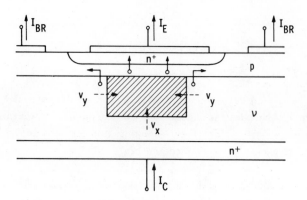

Fig. 22   Cross section showing the dynamic behavior of the current-induced base during the CIB storage time.

4.2 Waveforms and Profiles

Various current and voltage waveforms of interest are sketched in Fig. 23. It is assumed here that there is no snubber network to limit the rate of V_{CE} increase. The storage time is usually defined with respect to the $I_C(t)$ or $V_{CE}(t)$ waveforms, for this discussion we can approximate the storage time t_s by the time required to remove charge from regions 2 and 3 (see Fig. 21). We can further subdivide t_s as

$$t_s = t_{s',RB} + t_{s',CIB} \qquad (7)$$

where the first term on the right of this equation is the "remote-base" storage time and is the time required to remove charge from region 3. The second term corresponds to the time for charge to be removed from the current-induced base. As indicated in Fig. 22, the edges of the current-induced base will move with velocities, v_x and v_y which will depend on the reverse-base current I_{BR}, the amount of charge initially stored, and also the device geometry and impurity profile.[17]

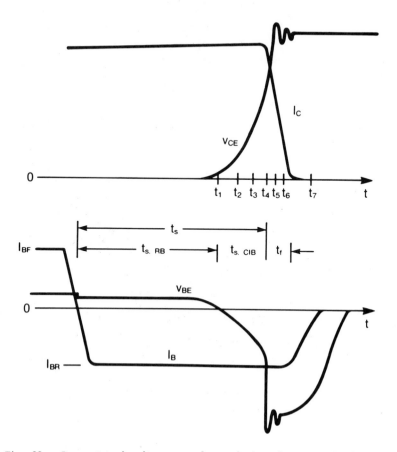

Fig. 23   Current and voltage waveforms during the turn-off interval.

The lateral velocity can be estimated by setting I_{BR} equal to the total number of holes "uncovered" by the moving boundary of the CIB. That is

$$I_{BR} = qZQ(t)v_y \tag{8}$$

where Z is the emitter perimeter and Q(t) is the excess charge per unit area in the CIB.

The field profile in the collector undergoes a rather dramatic change during the t_s interval. Consider the times t_1 through $t_{s,CIB}$. During this period the voltage $V_{CE}(t)$

is determined largely by the current density in the "unmodulated" portion of the collector together with its length W_C-W_{CIB}. Fig. 24 shows a plot of the electric-field and excess-carrier profiles in the collector for this time interval.

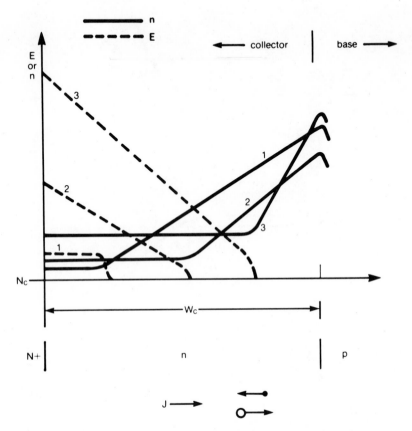

Fig. 24   Sketch of the electric-field and electron-density profiles for times t_1, t_2, and t_3 of Fig. 23. Conventional current flow and hole and electron flow are indicated at the bottom of the diagram.

As the lateral dimension of the CIB shrinks to a small value, the carrier density begins to increase to large values. That is

$$J = I_C/Zy(t) = 2qD_C\Delta n/W_{CIB} \tag{9}$$

where Δn is the excess-electron concentration at the CB junction. Quasi-neutrality will be reached in the CIB approximately within a transit time of this region. As v_y begins to increase, a situation will arise where the increase in electron density required by equation (9) can no longer be effectively neutralized. That is, $n > p + N_C$ will occur within the CIB, and a significant electron negative space charge will begin to build up. It is believed that this condition precedes blocking of the base-collector junction, and at this point the fall-time interval begins.

4.3 Avalanche Effects

The field profiles for times in the t_f interval are sketched in Fig. 25. During this period the profile slope changes because the current density is decreasing and the number of mobile carriers is no longer sufficient to affect the shape of the profile.

If the peak field reaches a value $E_c \cong 1 \times 10^5$ V/cm, which denotes the regions where significant avalanche generation begins, holes can be generated near the n^+ contact

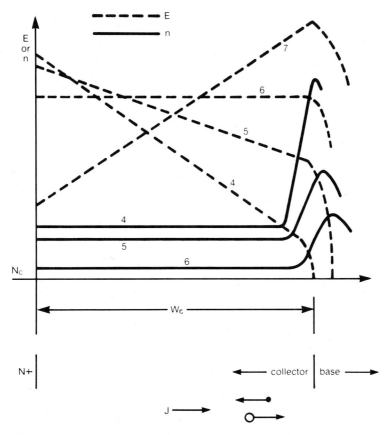

Fig. 25   Sketch of the electric-field and electron-density profiles for times t_4 through t_7 of Fig. 23. Conventional current flow and hole and electron flow are indicated at the bottom of the diagram.

by impact ionization. This hole current will drift toward the metallurgical base and can either flow out the base terminal or be injected into the emitter.

Figs. 24 and 25 show that there are two situations for concern. The first occurs when the peak field is located at the n-n^+ junction. In this case there is still a CIB and most of the avalanche-generated holes will be injected into the emitter. If the product of this avalanche hole current and the (intrinsic) gain of the transistor is comparable to the collector current, the transistor will begin to turn on. The resulting increase in current density will enhance the field at the n-n^+ junction and at the same time reduce V_{CE}, a runaway situation referred to as "avalanche injection".[2]

Avalanche injection will not occur if the collector-base junction begins to block before the peak field reaches E_c. In this case the field profiles follow the pattern shown in Fig. 25 and significant hole current can begin to flow out the base terminal, a situation sketched in Fig. 26. As indicated, there is a lateral field set up in the metallurgical base which tends to turn on the center of the emitter. Again a runaway situation can develop which we refer to as "avalanche instability" since it is similar to the thermal instability which leads to hot-spot formation.

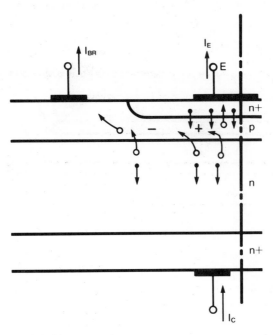

Fig. 26   Cross section showing the flow of holes and electrons generated by avalanche near the collector-base junction.

Therefore there are two possibilities which lead to a negative resistance or runaway situation during inductive turn-off. These are avalanche injection and avalanche instability. Both mechanisms culminate in second breakdown, which leads to a further localization of the plasma. Unless measures are taken to rapidly bypass the collector current[18] destructive alloying of the collector-emitter contacts will occur.

4.4 Reverse Safe Operating Area

During turn-off the I_C-V_{CE} locus must stay within a region defined as the Reverse Safe Operating Area (RSOA) if second breakdown is to be avoided. Nondestructive measurements of the onset of second breakdown show that the general shape of the locus is as indicated in Fig. 27. Although there is currently no satisfactory design theory for predicting this locus, it is in qualitative agreement with the arguments advanced in this paper.

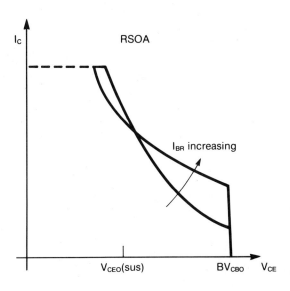

Fig. 27   Sketch of the Reverse Safe Operating Area showing the influence of I_{BR}.

As I_C is further decreased a point is reached where the collector-base junction begins to block before E_C is reached at the n-n$^+$ substrate. In this case the field profile will peak at the collector-base junction and second breakdown will be triggered by avalanche instability. When avalanche instability is the triggering mechanism, increasing I_{BR} leads to an <u>increase</u> in the safe collector current as indicated in Fig. 27. In current practise, advantage is taken of this portion of the RSOA by using a snubber network.

5. CONCLUSION

Various limits and capabilities of bipolar transistors have been described. When compared with the MOSFET and GTO it is concluded that all three devices have a place within the power-electronics spectrum.

There are a number of trends that are discernible within the bipolar industry. There is a tendency towards larger devices using thyristor packaging techniques. Also, users are becoming more aware of the advantages to be gained by controlling the switching locus and using more sophisticated base-drive techniques. Finally, the fact that the turn-off behavior is only partially understood and characterized means that we can expect improvements in the design and testing of bipolar transistors that will increase their performance beyond the limits described in this paper.

ACKNOWLEDGEMENT

The work on large-area transistors was carried out at Westinghouse Research and Development Center, Pittsburgh, Pa. and was supported by a contract from NASA-Lewis Research Center, Cleveland, Ohio. Thanks are due D.L. Blackburn, T.M.S. Heng, Y.C. Kao, E. Simon, and G.S. Sundberg for helpful comments and suggestions.

REFERENCES

1. R.M. Scarlett and W. Shockley, "Secondary Breakdown and Hot Spots in Power Transistors," IEEE Int. Conv. Rec., 11, Part III (1963) 3-13.

2. P.L. Hower and V.G.K. Reddi, "Avalanche Injection and Second Breakdown in Transistors," IEEE Trans. Electron Devices, ED-17 (April 1970) 320-355.

3. L.E. Clark, "High Current-Density Beta Diminution," IEEE Trans. Electron Devices, ED-17 (September 1970) 661-666.

4. J. Olmstead et al., "High-Level Current Gain in Bipolar Power Transistors," RCA Rev. 32 (1971) 222-246.

5. Y. Nakatani et al., "A New Ultra-High Speed High-Voltage Switching Transistor," Proc. Powercon 7, Paper J3, San Diego, 1980.

6. T.C. New et al., "High Power Gate-Controlled Switch," IEEE Trans. Electron Devices, ED-17 (September 1970) 706-710.

7. H.W. Becke and J.M. Neilson, "A New Approach to the Design of Gate Turn-Off Thyristors," IEEE Power Elec. Spec. Conf., Los Angeles, 1975, 292-299.

8. P.L. Hower, "A Comparison of Bipolar and Field-Effect Transistors as Power Switches," IEEE Ind. Appl. Soc. Conf. Record-1980, Cinncinatti, 682-688.

9. P.L. Hower, "Application of a Charge-Control Model to High-Voltage Power Transistors," IEEE Trans. Electron Devices, ED-23 (August 1976) 863-870.

10. H. Schlangenotto and W. Gerlach, "On the Effective Carrier Lifetime in p-s-n Rectifiers at High Injection Level," Solid-State Elec., 12 (1969) 267-275.

11. N. Oh-uchi et al., "A New Silicon Heterojunction Transistor Using Doped SIPOS," IEDM Tech. Digest (1979) Washington, 522-524.

12. P.L. Hower, "Optimum Design of Power Transistor Switches," IEEE Trans. Electron Devices, ED-20 (April 1973) 462-435.

13. P.L. Hower and W.G. Einthoven, "Emitter Current-Crowding in High-Voltage Transistors," IEEE Trans. Electron Devices, ED-25 (April 1973) 465-471.

14. P.L. Hower, "A New Method of Characterizing the Switching Performance of Power Transistors," IEEE Ind. App. Soc. Conf. Record-1978, Toronto, 1978, 1044-1049.

15. K.S. Tarneja and P.L. Hower, "A New High Power Transistor," IEEE Ind. App. Soc. Conf. Record-1981, Philadelphia.

16. E.D. Wolley, "Gate Turn-Off in p-n-p-n Devices," IEEE Trans. Electron Devices, ED-13 (July 1966) 590-597.

17. P.L. Hower, "A Model for Turn-Off in Bipolar Transistors," IEDM Tech. Digest (1980) Washington, 289-292.

18. D.L. Blackburn and D.W. Berning, "Some Effects of Base Current on Transistor Switching and Reverse-Bias Second Breakdown," IEDM Technical Digest (1978) Washington, 671-675.

DISCUSSION
(Chairman: Prof. P. Leturcq, Institut Nationale des Sciences Appliquées, Toulouse)

M. Adler (General Electric, Schenectady)
You commented that you expect the bipolar speed to increase, presumably at fixed turn-off gains. What factors would be responsible for that?

P. Hower (Unitrode Corp., Watertown)
The structure I assumed here was actually the 30 mill-wide fingers which is fairly crude for a bipolar. If you reduce the width of the fingers you can improve the shape of the rising part of the V_{CE} voltage and the device will turn off faster. It is simply the case that a finer geometry has superior characteristics.

D. Silber (AEG-Telefunken, Frankfurt)
I would like to draw attention to the reduction of the emitter Gummel number, which is similar to our experience with electron irradiated samples. From thyristor modelling I believe that it is not assumed that the emitter Gummel number is reduced.

P. Hower
This refers to the slide which showed G_E versus τ. If you electron irradiate to reduce τ, we have found that the emitter Gummel number goes down. The profiles shown by Dr. Adler, however, indicated that the carrier concentration on the sides were not reduced, and constant Gummel numbers were observed.

R. Sittig (BBC Brown, Boveri & Co. Ltd., Baden)
I think the important point is that Dr. Adler applied a constant current density. Therefore the carrier concentration cannot be related directly to the emitter characteristics.

POWER MOSFETS

J. TIHANYI
Siemens AG, Munich, Fed. Rep. Germany

SUMMARY

Novel power MOSFETs suitable for the power range below 10 kW are faster than bipolar power transistors and need less input current. Although easy to use, they also have some inherent problems. New advanced devices such as MOS thyristors, MOS Darlingtons, opto-MOS thyristors and triacs are now feasible in which MOS and bipolar structures are functionally integrated to good advantage.

1. INTRODUCTION

Since the introduction of genuine "power MOSFETs" about two years ago,[1,2] MOS power devices have found acceptance in the rapidly developing field of power electronics. There are today numerous suppliers[3,4,5] advertising devices operating at up to more than 30 A and providing blocking voltages up to 1000 V. Experience with power MOSFETs already shows them to be semiconductor devices with a highly promising future, especially in connection with microcomputer-controlled automatic systems and power systems. However, the past has also shown that certain problems exist that need to be solved before the advent of the power MOS era is possible. The present report is more concerned with discussing the difficulties and problems inherent in power MOS technology than the advantages it has to offer. Whereas earlier enthusiastic reports tended to convey the impression that the new power MOSFETs are omniscient electronic devices, the present author will try to be more realistic by describing the problems still to be solved before power MOSFETs can be endowed with omniscience.

A novel power MOSFET is more closely akin to an IC than to a classical power device such as a bipolar transistor or thyristor. As shown in Fig. 1, it contains tiny cells, is implemented in planar technology, and its interconnections are established by ultrasonic bonding. The cross section of a typical power MOSFET shown in Fig. 2 reveals its principal structural features: an epitaxial layer grown on a high-doped substrate, a metallic source covering the buried polysilicon gate, and cells produced by ion implantation, resulting in a short effective channel.

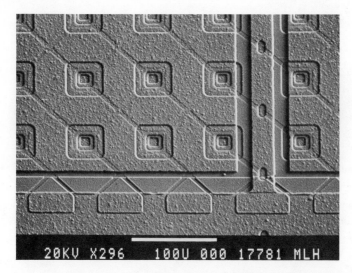

Fig. 1   Cell structure of a typical power MOSFET (SIPMOS BUZ 45).

All types of power MOSFET are fine structures, the least fine[2] having a minimum line-width of 5 µm. The process steps are largely identical with those used for the fabrication of n-channel MOS ICs.

At this point it may be noted that present experience shows IC technology to be a limiting factor: decreasing yield with increasing chip area limits the economically viable

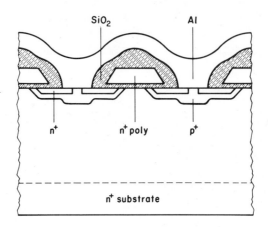

Fig. 2   Cross section of a power MOSFET.

size of chips. Besides this problem of relatively limited chip size there are also other physical parameters that need to be improved before the final breakthrough in power MOSFETs can be achieved.

2. FUNCTION OF POWER MOSFETS

Power MOSFETs are unipolar devices, in contrast to bipolar transistors and thyristors, in which both electrons and holes play a role. N-channel power MOSFETs are able to carry more current related to area because of the higher n-carrier (electron) mobility. It is for this reason that these are almost exclusively used.

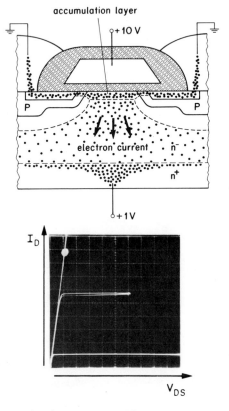

Fig. 3   Electron current in the on-state. Electrons are accumulated on the surface between the cells.

Fig. 3 shows the cross section of a power MOSFET cell in the on-state. The gate is positively biased. Since it is isolated, no input current flows. With positive gate bias, electrons accumulate at the surface and the highly conducting surface layer is the source for the ohmic electron current flowing vertically in the low-doped n^- drain region. In its on-state the device presents a resistance that depends mainly on the thickness and doping concentration of the n^- expitaxial layer. Current flows only between the cells below the gate. All cells are paralleled. In novel devices current flows in about 60 - 70 % of the total chip area. The resistance of the n^- layer rises with the temperature such that the current decreases in places that are at a higher temperature. This effect provides inherent temperature stability due to the absence of hot spots in MOSFETs.

Fig. 4 shows the off-state situation. At gate voltages below the threshold voltage the p-barrier region blocks the flow of electron current between the n^+ source and the n^- drain. At higher drain voltages the n^- epitaxial layer is partially depleted and the voltage drops in the space-charge layer. If the maximum electric field at the p^+n^- junction exceeds the threshold, breakdown occurs. Modern power MOSFETs, if properly designed, reach the bulk breakdown voltage given by the thickness and/or doping of the n^- epitaxial layer. The highest blocking voltage reported[6] is 1000 V. For high blocking voltage, a low-doped thick n^- epitaxial layer is needed, which means that a high-voltage device will always have a higher on-resistance than a low-voltage device, assuming a chip area of the same size. The relation between on-resistance R_{on} and blocking voltage V_B is approximately[7] $R_{on} = const \cdot V_B^{2.6}$.

This physical effect is the principal limiting factor of power-MOSFET technology and, as will be discussed in detail later, is an obstacle to the fabrication of high-current high-voltage devices.

Fig. 5 shows the current flow in the pentode region. The electrons flow along a narrow current path from the short channel to the drain. The current is conducted solely by electrons; the carrier concentration in the n^- epitaxial layer is equal to or higher than the dopant concentration. There is no stored charge. An exact description of the situation is only possible by two-dimensional numerical modelling,[8] which is rather complicated. It is of interest to note the depletion of the surface between the cells, which reduces the Miller capacitance and makes the high switching speed of power MOSFETs possible.

A very interesting operating mode is the reverse operation shown in Fig. 6. In the reverse mode the device acts like a power rectifier under forward bias. The epitaxial layer is swamped with carriers in the same way as bipolar devices in the on-state. Al-

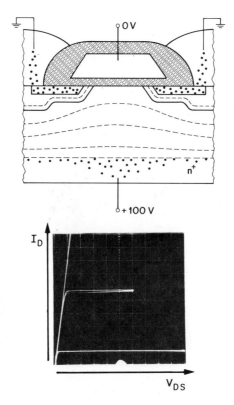

Fig. 4   The p-barrier region blocks electron flow in the off-state.

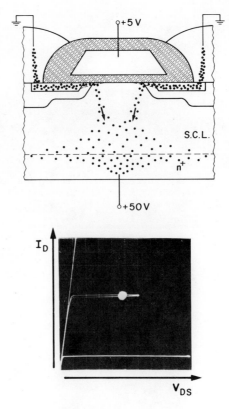

Fig. 5   Current flow in the pentode region. The surface between the source cells
         is depleted.

though the existence of a reverse rectifier in power MOSFETs could be a hidden bonus in many applications,[9] the large amount of stored charge can, as will be shown
later, reduce this benefit drastically.

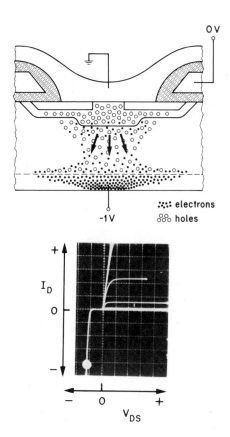

Fig. 6   The source cells act like p-i-n diodes at negative drain voltage.

The described function is common to all types of power MOSFET with vertical current
flow, independent of make and noncritical differences in the fabrication process. The
rival families of power MOSFETs differ from each other mainly in the percentage of
chip area used for current flow, in device-design cleverness with respect to the high-
voltage properties, and in the processing technology chosen in the interests of high
yield and low cost. It is to be hoped that power MOS prices will, by the mid-eighties,
equal the price of bipolar devices with the same chip area.

Although the described unipolar function leads to electrical parameters that are usually more favorable than those of bipolar devices, MOSFETs also have some handicaps. Two such handicaps are the relationship between R_{on} and V_B, which excludes high-current devices from the high-voltage range, and the limitation of chip size due to the use of IC fabrication technology. The favorable parameters are the following:

a) high working speed
   no input current needed

b) no storage charge
   built-in reverse diode

c) easy paralleling
   no thermal instabilities.

The handicaps noted earlier are offset in many applications by the advantages. For the future it is essential to determine the specific limitations of power MOS technology as a whole so that these may be overcome and the area of application expanded.

3. PERFORMANCE LIMITS

3.1 Current-Carrying Capability

The formula $R_{on} = const \cdot V_B^{2.6}$ defines the practical current-voltage range of power MOSFETs. Fig. 7 shows the calculated ideal conductance related to area as a function of the blocking voltage in comparison to that of available MOS and bipolar devices. [10,11] Comparing MOS and bipolar devices at room temperature is, however, not entirely fair because it fails to take into account realistic operating conditions. The comparison shown in Fig. 8 was made under more realistic conditions: it was assumed that T_{case} = 75°C, T_j = 150°C and that the devices in a TO 3 package have a thermal resistance of about 1°C/W. The solid curve shows the maximum permissible direct current as a function of the blocking voltage for a MOSFET with a chip area of 0.5 cm^2 (the largest chip area currently conceivable with IC technology). Fig. 8 shows that even with a MOSFET larger than any which now exist, it is not possible to reach the average direct current of available bipolar power transistors. However, no existing bipolar device is able to operate at above 500 V, whereas 1000-V MOSFET's for currents in excess of 3 A are available. No one as yet has devised ways and means of implementing power MOSFETs for higher currents than bipolars.

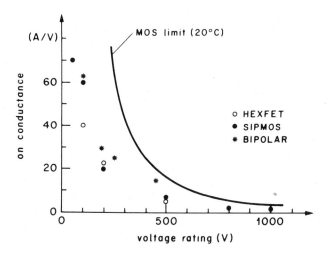

Fig. 7   Comparison of MOS and bipolar transistor on-conductances at 20°C.

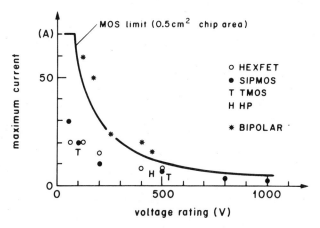

Fig. 8   Comparison of MOS and bipolar current capabilities under realistic conditions: T_j = 150°C, T_{case} = 80°C, R_{th} = 1°C/W (TO 3).

3.2 Switching Speed

The switching speed of power MOSFETs is determined by the parasitic capacitances shown in Fig. 9.[12] All the capacitances in the equivalent circuit must be charged and

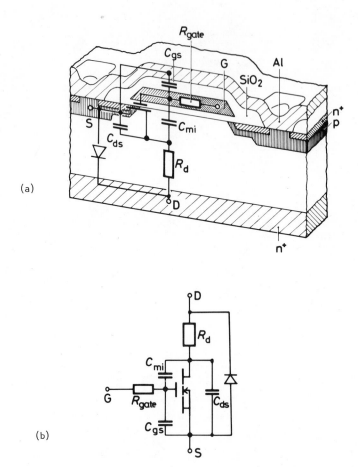

Fig. 9 a) Parasitic capacitances in a power MOSFET.
      b) Equivalent circuit with an integrated reverse rectifier.

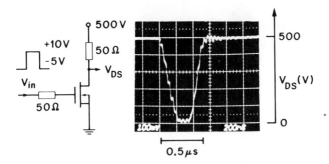

Fig. 10   A 500-V device with a chip area of 16 mm^2 switches 10 A·500 V power.

discharged during each switching operation. Neither the capacitances nor the switching times are influenced by the temperature, this being a big advantage over bipolar transistors. A further bonus offered by MOSFETs is their high speed. The switching times depend only on the input current charging the input capacitance of $C_{in} = C_{gs} + g_{mi} \cdot C_{mi}$, assuming a given chip size. As shown in Fig. 10, a 4 mm x 4 mm 500 V

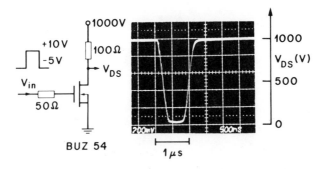

Fig. 11   A 1000-V device with 36 mm^2 chip area switches 10 A · 1000 V power.

SIPMOS device driven by a 50-ohm pulse generator is capable of switching a 10 A ·
500 V power with a minimum pulsewidth of 0.5 μs. This speed is beyond the capability of any bipolar power transistor. A further capability of the MOSFET is shown in
Fig. 11 which depicts the waveforms of a BUZ 54 SIPMOS FET while switching 10 A ·
1000 V using a pulsewidth of 1 μs. Although the speed can be increased further by
using a low-resistance driver, this requires a higher driving current.

In reality the statement "MOSFETs need no input current" is only true in the case of
slow switching. For high-speed switching, the required charging currents are by no
means negligible. Fig. 12 shows the peak input current as a function of the rise and

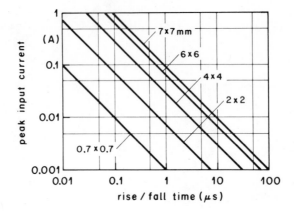

Fig. 12   Peak input current vs. rise/fall time.

fall times for various chip areas. As a result, the input current of MOSFETs is always smaller than that for bipolars, but with really low input currents only small MOS-
FETs are fast. In connection with the above discussion it may be added that "direct
microcomputer and IC compatibility" is only true for MOSFETs with the qualification
"at low working speed". Unlimited IC and microcomputer compatibility would exist only
with integrated input amplifiers as shown in a conceivable version in Fig. 13. Unfortunately, no one has as yet succeeded in realizing such integrated devices.

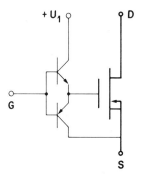

Fig. 13   An integrated input amplifier reduces the input current.

3.3 Charge-Storage Effects

Physically the current-transport mechanism in power MOSFETs is the carrier drift in an electric field. The current transport is ohmic and the carrier concentration does not exceed the doping concentration except in the narrow surface channel and the accumulation layer. Since the charge difference between the on and off states is much smaller than in bipolar devices, no "stored charge" exists in the bulk of MOSFETs as it does in bipolar transistors. However, there is an enormous excess charge in the surface accumulation layer, in the "Miller capacitance" under normal operating conditions.[12] The capacitances in the equivalent circuit, especially the Miller capacitance, are strongly voltage dependent as shown in Fig. 14. Under normal operating conditions an enforced overdrive is used in the input circuit to realize a short turn-on time. Thus the gate voltage is much higher at the end of the turn-on period than would be needed to switch to the smallest on-state resistance. The gate overvoltage charges the large Miller capacitance. On the other hand, when the MOSFET turns off, the excess charge has to be removed as shown in Fig. 15, before the current begins to decrease.

The charge stored in the overdriven input capacitance produces a time interval between the input-voltage drop and the output-voltage rise, which for the user is the same as the storage time known from bipolar transistors. Although the MOS storage time is much shorter (< 0.5 µs at 50 ohm input source resistance) than that of bipolar transistors (up to 5 µs at 10 A) and is not influenced by the load current or the temperature, MOSFETs do indeed have a storage time, but this can fortunately be reduced by the user by applying driving pulses of a special shape that is relatively easy to realize.

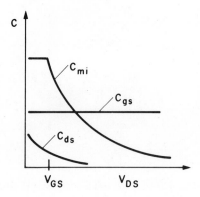

Fig. 14   Voltage dependence of parasitic capacitances.

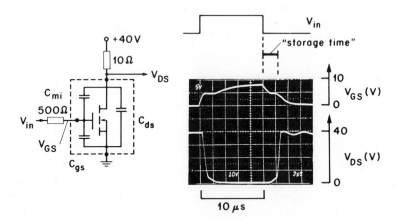

Fig. 15   Input and output waveforms of a power MOSFET inverter show a kind of "storage time".

Another storage effect influencing power-MOSFET applications is the charge storage in the reverse diode during inverse operation.[9] The inherent built-in reverse diode could be used very advantageously as a free-running diode in connection with inductive loads if it were faster and if its stored charge were smaller than that in currently available power MOSFETs. The problem is demonstrated in Fig. 16. The basic circuit

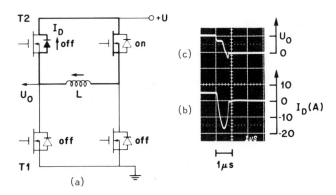

Fig. 16   Charge storage in the reverse rectifier.
a) Typical circuit and current flow in "free run".
b), c) Current and voltage waveforms.

configuration of inductively loaded MOSFET systems such as a.c. motor drives and relay drives is shown in Fig. 16a. When T_1 is on, the current in L rises. When T_1 is turned off the current flows through the reverse diode of T_2. When T_1 is turned on anew the stored charge in T_2 has to be discharged before the voltage drops at T_1 resulting in enormous stress from simultaneous high voltage and high current for T_1 as shown in Fig. 16b and c.

The current flowing through the inductance is only 5 A, but the peak current is more than 3 times 5 A for the relatively long period of about 1 µs. The stress caused by the current peak and the dissipated power during switching can, at pulse frequencies as low as 10 kHz, already exceed the total d.c. stress and dissipation. The situation is worse for high-voltage MOSFETs, which are only suitable for European network applications such as motor controls, which repesents perhaps the principal market for such devices. The user helps himself by employing various auxiliary circuits to reduce the peak current, but a more elegant solution would be to use a power MOSFET with a reverse diode specially tailored for motor-control applications.

3.4 Thermal Parameters

As is known from the literature, the unipolar current transport mechanism of power MOSFETs results in extreme thermal stability. There are, however, certain handicaps. Fig. 17 shows schematically the temperature response of the d.c. characteristics of

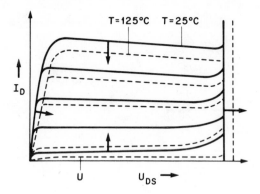

Fig. 17   Temperature dependence of power MOSFET characteristics.

power MOSFETs. The current actually decreases with increasing temperature in the triode region, and high currents in the pentode region result in extraordinary ruggedness due to the absence of hot spots. It is normal for a power MOSFET to be shorted for many microseconds at its peak voltage without degradation (Fig. 18).

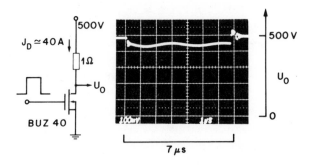

Fig. 18   Power MOSFETs can be shorted at the rated voltage.

At lower currents in the pentode region there is however a need for caution. The threshold voltage decreases with rising temperature, resulting in a rise in current. This effect must be taken into account when using power MOSFETs in class-A amplifiers or other analog applications. By biasing a power MOSFET with a constant current-constant voltage regime, the current flow will be shifted to the point of highest thermal resistance of the device, an effect that is liable to disturb instruments for measuring thermal resistance.

4. TRENDS

Power-MOSFET people seem to be aware of the foregoing problems and are trying to solve them with various approximations. There is a general desire to combine the favorable aspects of the MOS and bipolar devices while avoiding their drawbacks. Some innovative approaches are represented by the devices discussed below.

4.1 MOS Darlingtons

A new device shown in Fig. 19 is the MOS Darlington, which combines the low input current of MOSFETs with the higher current-carrying capability of bipolar power transistors. Their switching speed is, however, almost the same as that of bipolar Darlingtons.[13]

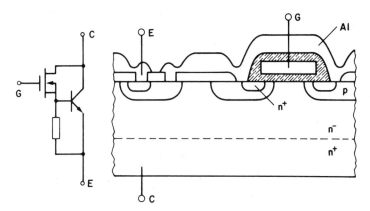

Fig. 19   Cross section and equivalent circuit of a MOS-Darlington.

4.2 MOS Thyristors and Triacs

Another approach is the light-fired MOS thyristor[14] and the MOS triac.[15] These consist essentially of a MOSFET with low input current combined with a vertical or lateral thyristor with effective emitter shorts. This shorting results in a good dV/dt, but a

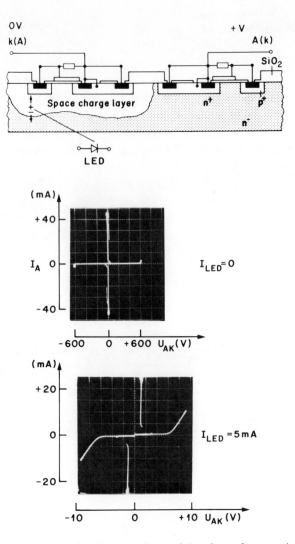

Fig. 20   Light-fired MOS-triac and its d.c. characteristic.

need for high base current. High base current can, however, be readily supplied by the integrated MOSFET, which needs only a small input power. Given appropriate design, current extension can be fully avoided, resulting in an excellent dI/dt and short turn-on time.

Since the light-fired thyristor needs practically no input current, this "no input current" could obviously be supplied by a small integrated photodetector if ways and means were found to integrate one with the other. Furthermore it is relatively easy to integrate two light-fired thyristors in antiparallel configuration on a single chip to obtain a light-activated triac.

Such a triac structure with its d.c. characteristic is shown schematically in Fig. 20. Many of its parameters are superior to those of comparable bipolar devices: the input sensitivity is higher, dV/dt, dI/dt, and commutation dI/dt are all superior, the price is lower than that of conventional devices, and there is the possibility of integrating a zero-crossing capability to realize a complete solid-state relay function in a single chip.

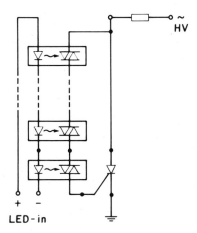

Fig. 21   MOS-triacs as the firing circuit for high-power high-voltage thyristors.

The first device planned is a 0.3 A/600 V relay which can also second-source the rare BOD's. These relays will greatly stimulate the application of large power thyristors because they can be connected in series to realize high breakdown voltages as shown in Fig. 21. Further developments in MOS thyristors and MOS opto-devices are highly probable.

4.3 Feasibility and Demand

Besides the two functionally integrated MOS and bipolar structures described above, other devices are conceivable and also needed. These will be switches rather than thyristor-like structures. The significant features of bipolar and MOS power switches are listed in Table 1 to permit comparison.

TABLE 1

Structure	Input	Speed	Current	Reverse Diode
Bipolar	-	-	+	--
MOS	+	+	-	-

Table 2 lists on a subjective basis the parameters of prospective new devices for which there is a definite demand. The motor switch, which will satisfy the demand for efficient motor control, is a simple 0.5 ohm/400 - 500 V power MOSFET with an improved reverse diode. The reverse stored charge must be sufficiently small to allow an operating frequency of about 20 kHz to 25 kHz. The second device, a "high-voltage switch", will have the same input current and speed as an MOS device, but a higher current-carrying capability. It will not fall below R_{on} = const $\cdot$ $V_B^{2.6}$, and will make it possible to realize high-voltage high-speed circuits with higher currents than at present.

The "super motor switch" will possibly resemble the motor switch, but will have a considerably larger chip area of some cm^2. It will need a MOSFET technology that is basically different from the present. The term "microcomputer-compatible power switch" implies a device such as a normal MOSFET with an improved reverse diode and an input amplifier that allows it to be driven directly from microcomputer outputs and to realize high switching speeds. Finally, the term "large switch" implies a utopian device with a microcomputer-compatible input, MOS switching speed, a very large chip area of perhaps 10 cm x 10 cm, and a very good reverse diode. This device would be a MOSFET version of today's large thyristors.

TABLE 2

Structure	Input	Speed	Current	Reverse Diode
Motor switch	+	+	-	+
HV switch	+	+	+	-
Super motor switch	+	+	+	+
Microcomputer-compatible switch	++	+	+	+
Large switch	++	+	++	+

In our view, only the MOSFET with an improved reverse diode and the large MOSFET with microcomputer-compatible input are really feasible at the present time, while all the other devices belong to the more distant future.

5. CONCLUSIONS

Novel power MOSFETs are advanced power-electronics devices for applications below the 10-kW power level. Their low input power and high switching speed simplifies old-established power systems and opens the way to new possibilities. Some feasible developments could expand the 10-kW power range in the near future and open the way to new approaches for motor-control systems. Power-MOSFET technology is a starting point for new functionally integrated MOS bipolar devices with emphasis on light-fired thyristors and triacs. Power-MOSFET technology will lead to genuinely microcomputer-compatible power switches.

REFERENCES

1. "The HEXFET: A New High in Power MOS," Electronics Design, June 7, 1979.

2. "The SIPMOS Device Family," Press conference, February 1980, Munich.

3. "TMOS Data Sheet," Motorola, October 1980, Phoenix.

4. "HP Power MOSFET," Press information, Hewlett-Packard, October 24, 1980, Palo Alto.

5. "Power MOSFET Data Sheet," Intersil, September 1980.

6. J. Stengl, H. Strack, and J. Tihanyi, "1000 V 2-Ohm Power MOSFET," in preparation.

7. C. Hu, "A Parametric Study of Power MOSFETs," IEEE Dev. Res. Conf., 1979, Denver.

8. A.W. Wider, C. Werner, and J. Tihanyi, "2-D Analysis of the Negative Resistance Region of Vertical Power MOSFETs," IEDM (1980) Washington, 95-99.

9. B. Pelly, B. Fragale, and B. Smith, "The HEXFET's Integral Reverse Rectifier - A Hidden Bonus for the Circuit Designer," IR Application Notes AN-934, 1980.

10. A. Lidow, T. Herman, and H.W. Collins, "Power MOSFET Technology," IEDM (1979) Washington, 79-83.

11. J. Tihanyi, "MOS Power Devices - Trends and Results," ESSDERC (1980), York.

12. J. Tihanyi, P. Huber, and J.P. Stengl, "Switching Performance of SIPMOS Transistors," Siemens Res. and Dev. Reps., 9 (1980) 195-199.

13. B. Boss, "Power Transistors Unite MOS, Bipolar," Electronics, April 21, 1981.

14. L. Leipold, W. Baumgartner, W. Ladenhauf, and J.P. Stengl, "A FET-Controlled Thyristor in SIPMOS Technology," IEDM (1980) Washington, 79-82.

15. J. Tihanyi, "Functional Integration of Power MOS and Bipolar Devices," IEDM (1980) Washington, 75-78.

DISCUSSION

(Chairman: Dr. H. Becke, Bell Laboratories, Murray Hill)

W. Zimmermann (BBC Brown, Boveri & Co. Ltd., Baden)

Do you believe that it is possible with the BUZ-54, which is a 1000-V device, to construct a motor control, taking into consideration the fact that there are problems with reduced switching time and high losses in the absence of a snubber circuit?

J. Tihanyi (Siemens AG, Munich)

I believe that it can be done. We hope that with more precise techniques the reverse diode can be made about one order of magnitude faster so that the switching speed of the transistor would be the determining factor and not the diode itself.

A. Marek (Brown Boveri Research Center, Baden)

I think in most applications it is not necessary to wait for the diode because usually by independently controlling both switches you can introduce a time delay between switching off the lower one and switching on the upper one.

J. Tihanyi

This can be done but the circuit remains simple only if you need no other components than the MOS-FET, and that's the goal we want to achieve.

EPI AND SCHOTTKY DIODES

R.J. GROVER
Mullard, Hazel Grove, England

SUMMARY

Power can be efficiently converted from the mains to low voltage, high current, d.c. in the switched mode power supply. The output stages require low voltage, fast diodes to minimize the losses. Reverse recovery times of 30 ns, and reverse voltages of 200 V can be achieved by replacing the double-diffused structure with an epitaxial one. The well-controlled, narrow base width allows low stored charge to be achieved, together with low forward voltage drop. This technology is well established in the range 1 - 70 A. An extension of the technique to higher voltage is being studied, and it appears that it can be used with advantage at 400 V. Alternative lifetime killers to gold may give benefits, e.g. in the hot leakage current. A particularly difficult application is that of antiparallel diode to a power MOS switch, because of the very low forward voltage needed to avoid charge storage occurring in the switch.

For output voltages of < 12 V the forward losses of epi diodes are too great, and the Schottky diode is used. In these devices there is no charge storage, so switching times are only dependent on the capacitance of the device and the impedance in the circuit. The absence of conductivity modulation means that the epitaxial layer introduces a series resistance into the device. This limits the device to thin epi and, hence, low reverse voltage, < 60 V.

The barrier height can be varied by choosing the barrier metal, and values of 0.5 - 0.9 eV have been used. It can be shown that values of 0.6 - 0.7 eV are optimum for typical applications. Molybdenum is thus a good choice for the barrier metal. The reverse voltage breakdown of the device can be optimized, with a simple structure, using 2-D computer calculations of the field distribution at the edge of the device. The future requirements of even lower output voltages ($\sim$ 2 V), to power advanced logic circuits, may need lower barrier height devices. In addition, higher frequencies may demand lower capacitance. Schottky devices are readily available for 1 to 70 A operation. Together with epi diodes, there seems to be no basic reason why higher current operation cannot be achieved. However, the need for such devices is not apparent. This is probably because the development of the other components, e.g. power switches and transformers, is lagging behind. Also, the requirement for output currents of > 100 A at low d.c. voltages seems limited.

1. EPI DIODES

1.1 Introduction

Epi diodes are silicon rectifiers made using epitaxial material, as opposed to the conventional double-diffused types which use bulk silicon slices. Good control of narrow ($\sim$ 10 μm) base widths is achieved and, in conjunction with gold killing, it is possible to produce very fast reverse recovery times and low forward voltage drops. These characteristics make the devices the best choice for operation in switched mode power supplies (SMPS) with output voltages of 12 - 50 V d.c., and output currents of 1 - 100 A. The breakdown voltage achieved (about 200 V) limits the application to output voltages of 50 V in a typical SMPS design.

1.2 Basic Device Concepts

In forward bias a simple model of the behavior of the p^+-n-n$^+$ structure used is as follows. Under forward bias holes are injected from the p^+ anode region into the n-base. They diffuse across the base, recombining in time with lifetime τ and in distance with diffusion length $L_p = \sqrt{D_p \cdot \tau}$, where D_p is the hole diffusivity, related to the hole mobility μ_p by Einstein's relation

$$D_p = \frac{kT}{q} \mu_p$$

Once the holes have recombined, the current has to be carried by the majority carriers in the base. This electron current gives rise to an ohmic voltage drop, dependent on the resistivity of the n-type base region. This extra contribution to the forward voltage across the diode can be minimized, for a given lifetime, by having a base width as small as possible. Approximately, the forward voltage V_F will be that of the ideal diode, if the diffusion length L_p is much greater than the base width. This ideal diode V_F is

$$V_F = \frac{kT}{q} \log_e \frac{I_F}{I_o} \quad , \quad \text{for } V_F > \frac{3kT}{q}$$

where I_o is a constant, and I_F is the forward current. For values of $L_p < W$ the V_F will rise rapidly as the current flow mechanism becomes increasingly that of electrons drifting across the base. This is shown schematically in Fig. 1.

In practice the diodes operate in high-injection conditions, that is the injected-hole density is greater than the electron density due to the base doping. In this situation

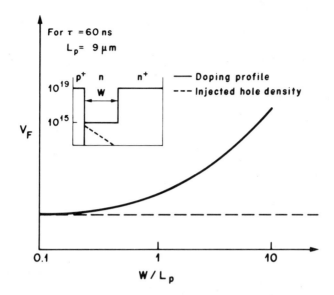

Fig. 1   Simple model of diode in forward bias (low injection) V_F as a function of the ratio of base width W to diffusion length Lp.

the device behaves like a p-i-n diode, and injection of both holes and electrons into the base occurs. Thus the base resistivity is modulated to a lower value. This reduces the V_F obtained for a given L_p (< W) compared with the simple low-injection model. Calculation of the V_F under these conditions is complex, as shown in Fig. 2.

Fig. 2 shows the results of computer calculations for narrow base diodes of low lifetime.[1] The model, which follows the work of several authors[2,3,4,5] makes the following assumptions:

a)   One-dimensional structure

b)   Homogeneously doped regions

c)   Quasi fermi levels are flat across the space-charge layers with no recombination in these layers

d)   High injection in the base, low elsewhere

The effect of the anode and cathode regions is described by 's' and 'h' parameters. In the case of the p⁺ anode diffusion, for example, the electron current in the base next

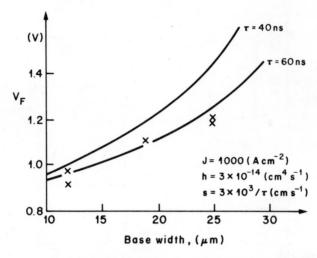

Fig. 2   Complex model of diode in forward bias (high injection)
            V_F against lifetime τ and base width W.

to the p^+ - n junction, J_{no}, is given by

$$J_{no} = qs_1 \, n_o + qh_1 \, n_o^2$$

where s_1 = anode s parameter
      h_1 = anode h parameter
      n_o = the concentration of carriers at the p^+ - n junc-
               tion

A similar expression holds for the cathode parameters s_2, h_2. A value of 3 x 10^{-14} cm^4/s has been chosen for h_1 = h_2, and a value of s_1 = s_2 = 3 x 10^3/τ cm/s where τ is the base lifetime in microseconds.

Comparison of a large number of V_F, Q_s and t_{rr} results and theory suggest that these values are correct. The experimental results for the normal epi diodes, which are gold killed, fit reasonably well with the V_F for a lifetime of 60 ns.

In reverse bias the depletion region extends into the lightly doped n-type material. Assuming step profiles for the p^+ and n^+ regions, the breakdown voltage V_{BD} is given by

$$V_{BD} = \frac{\varepsilon_s \varepsilon_o E_c^2}{2 N_D}$$

where E_c = critical field for breakdown

N_D = doping level in the base

ε_s = dielectric constant of Si

ε_o = permittivity of free space

If the doping level is low, however, the depletion region reaches the n^+ substrate, and in this case the breakdown voltage is given by

$$V_{BD} = E_c W - \frac{q N_D W^2}{2 \varepsilon_s \varepsilon_o}$$

A more exact calculation of the breakdown voltage takes into account the profiles of the p^+ and n^+ regions, and replaces the critical field assumption with a calculation of the avalanche integral. This reaches unity at breakdown. This 1-dimensional calculation is shown[6] in Fig. 3.

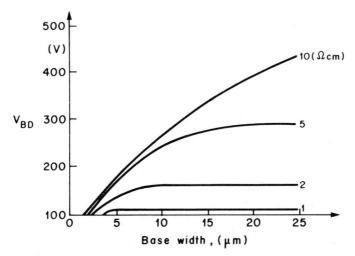

Fig. 3   Variation of breakdown voltage V_{BD} with base width W and resistivity as parameter. One-dimensional avalanche integral calculation.

The switching from forward to reverse bias is limited by the time taken to extract the charge stored in the diode. For rapid switching the lifetime in the base, τ, has to be low. Given the low lifetime and low forward voltage requirements, the base width has to be kept to a minimum. This limits the breakdown voltage of the device.

1.3 Typical Properties

The devices in production have properties as shown in Table 1. Similar devices are made by several other manufacturers. The devices achieve a forward voltage drop of 0.85 - 0.90 V at the working current. Blocking capability is up to 200 V. The turn-off behavior of the devices can be described in terms of the reverse recovery time, or the stored charge, Q_s (Fig. 4).

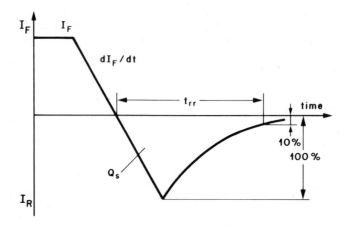

Fig. 4   Definition of t_{rr} and Q_s.

The devices turn-off in 25 - 60 ns under the conditions quoted. The apparent increase in t_{rr} with increasing device size is a result of the fixed measurement condition. If the devices were measured under conditions where the current density, and rate of change of current density were the same, then they would have the same t_{rr}. The recovery time and stored charge vary with measurement conditions in the following way:

a)   Higher forward currents (I_F) increase both t_{rr} and Q_s
b)   Lower reverse voltages increase both t_{rr} and Q_s
c)   Higher rates of fall of I_F reduce t_{rr}, but increase Q_s
d)   Higher junction temperatures increase both t_{rr} and Q_s

TABLE 1

Epi Diode Range

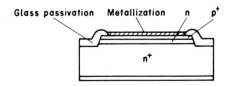

Chip details

	$I_{F(AV)}$	Max V_F at $I_{F(AV)}$	Max t_{rr}	Max Q_s	Max I_R
	1.3	0.80	25	15	0.01
	3.5	0.85	30	20	0.01
	7	0.90	35	15	0.6
	2 x 10	0.90	35	15	0.6
	14	0.90	35	15	1.3
	23	0.85	50	20	2.5
	30	0.90	50	20	2.5
	56	0.88	60	35	5.0
	70	0.85	60	50	7.0
Units	A	V	ns	nC	mA
Notes		1	2	3	1

General characteristics

V_{RRM} = 50/100/150/200 V

T_j max = 150°C

Notes: 1) At 100°C, at V_{RRM}

2) At I_F = 1 A to V_R = 30 V at dI/dt = - 50 A/μs

3) At I_F = 2 A to V_R = 30 V at dI/dt = - 20 A/μs

It is important, therefore, when comparing similar devices, to have the same measurement conditions.

1.4 Suitability for Application

These devices are well suited to rectification at frequencies of 20 kHz or more, where the low forward losses and switching losses result in efficient operation. A common application is the switched mode power supply. One version of this is the forward converter shown in Fig. 5.

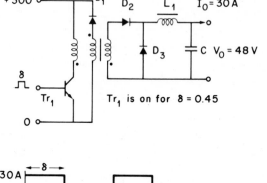

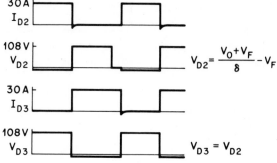

$$V_{D2} = \frac{V_O + V_F}{\delta} - V_F$$

$$V_{D3} = V_{D2}$$

Efficiency of $D_2 + D_3$: $\eta = \dfrac{P_O}{P_O + P_{D_2 + D_3}}$;

Switching losses are negligible $\rightarrow \eta = 98.5\ \%$

$P_O = 30$ A x 48 V = 1440 W

$P_{D_2 + D_3} = 30$ A x 0.75 V = 22.5 W

Fig. 5   Forward converter using epi-diodes at 20 kHz.

This is a simplified analysis, which neglects the variation in duty cycle used to compensate for input voltage and load variations. For a more complete description see Ref. 7. It can be seen that only 1.5 % of the output power is wasted in the diodes. The diode switching losses are negligible, and remain so to at least 100 kHz. The fast switching also reduces the stress on the transistor. When this turns on, D_3 changes from forward to reverse bias. During the reverse recovery period it is effectively a short circuit, causing high dissipation in the transistor.

1.5 Recent Progress and Results

This successful technology is now being extended to higher voltages, using thicker epitaxial layers. At 400 V there is direct competition with the double-diffused types. Fig. 6 compares results from these two approaches.

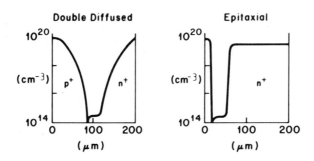

Technology	V_F	Q_s	t_{rr}
Double Diffused	1.55	80	65
Epitaxial 1	1.09	55	62
Epitaxial 2	1.25	23	46
Units	V	nC	ns
	150A	2A, 20A/μs	1A, 50A/μs

Active area = 0.02 cm^2; T = 25°C

Fig. 6   Comparison of 400-V technologies.

The devices made using epitaxial silicon have lower V_F, and are faster. This is in spite of similar base widths. The higher V_F and stored charge of the double-diffused types arise from the deep diffusions, which have low doping gradients near the base. This results in a widening of the base under high-injection conditions, and storage of charge in the end regions as well as the base.

The reverse recovery times of the existing devices have been reduced by diffusing in more gold. This results in lower Q_s and t_{rr}, but increases the hot leakage current due to bulk generation (Fig. 7).

Lifetime Process	Q_s	t_{rr}	I_R	V_F
*Gold T°C	5.0	25	200	1.00
Gold T+20°C	1.05	23	500	1.13
Gold T+40°C	0.90	21	500	1.39
Platinum T-40°C	5.7	26	24	1.08
Platinum T°C	2.9	23	29	1.10
	nC	ns	µA	V
	2A, 20A/µs	1A, 50A/µs	100°C, 200V	20A, 25°C

Active area = 0.02 cm^2

*Standard process

Fig. 7   Effects of various lifetime-controlling processes.

The forward voltage also increases substantially. An alternative center is platinum. This is already used on the smallest devices, BYV27/28. The results in Fig. 7 show that slightly higher V_F, for a given value of Q_s, is obtained, but that the hot leakage is greatly reduced. However, it has been stated[8,9] that lifetime processes should be compared using the switching and forward voltage characteristics measured at equal, practical current densities. Thus, plotting V_F at 10 A against Q_s at 10 A, 50 A/µs gives Fig. 8.

Here the platinum and gold processes give very similar results. The apparent improvement for platinum arises from the injection-level dependence of the lifetime. Thus at low levels the lifetime is longer, giving a higher Q_s, while at high levels the lifetime is shorter, giving a higher V_F. By measuring at equal, high-current levels the same lifetime applies to both characteristics. The lifetime dependence on current

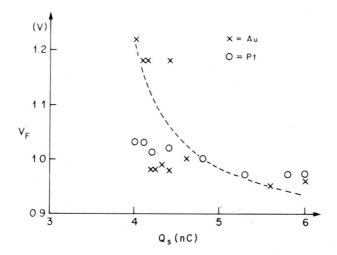

Fig. 8   Forward voltage at 10 A against Q_s at 10 A and 50 A/µs, for Au and Pt.

level is due to the shallowness of the platinum center (0.23 eV below the conduction band edge). This shallowness is also responsible for the greatly reduced leakage current.

1.6 Goals for Future Development

The production of thicker, high-quality epitaxial layers is being established. The use of layers up to 100 µm thick will be determined largely by economic considerations. At 100 µm the advantages of an epitaxial diode, compared with a double-diffused type, are reduced, and are offset by the increased cost of the slice. However, for voltages of 400 to 600 V, it seems clear that epi diodes will be introduced because of their superior performance. A 400-V version of the BYW29 is now being developed.

Platinum killing gives lower hot leakage current, and may allow higher maximum junction temperatures to be specified because of the improved thermal stability margin. However, increasing T_j max from 150°C to 175°C would put more stress on the passivation and package, and may be limited by this. The current passivation process, using glass, gives excellent stability at 150°C, at 200 V d.c. for 1000 hours. How-

ever, POLYDOX (=SIPOS) passivation has been shown to be more stable, and may be
necessary for 175°C blocking.

The economical use of this technology is being pursued in two particular ways. Firstly, the use of cheaper plastic packages, such as T0220, T0238. Secondly, the integration of diodes in one package. An example is a new diode now in development, which
will be a double, common cathode device in T0220, rated at 2 x 15 A.

1.7 Open Problem

One problem in the development and characterization of these very fast devices is the
measurement of lifetime and the reverse recovery parameters t_{rr} and Q_s. Techniques
like OCVD, ramp or step recovery are subject to large errors for lifetime of 60 ns.[10]

The measurement of t_{rr} and Q_s of these power devices is complicated by the capacitance of the device. This, together with small amounts of stray inductance, can produce ringing effects as the device turns off. Thus even Schottky diodes, which have
no minority carrier storage, but considerable capacitance, may appear to have Q_s and
t_{rr} values similar to epi diodes (Fig. 9).

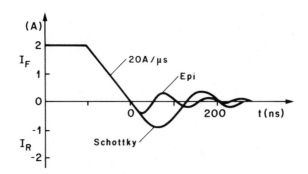

Fig. 9   Comparison of turn-off measured on:
              (a) An epi diode killed with Au
              (b) A Schottky diode of the same area

A capacitance of 100 pF, in series with a piece of wire, of length 2.5 cm and diameter
1 mm, is a resonant circuit with a time constant of 10 ns. However, this demonstrates
that the use of devices in applications, where t_{rr} < 20 ns is required, will present
considerable difficulties.

A second problem is the design of a fast diode for antiparallel use with VMOS devices. The built-in diode in a VMOS device comprises the p-body diffusion and the n on n^+ epitaxial layer (Fig. 10).

(a)   (b)   (c)

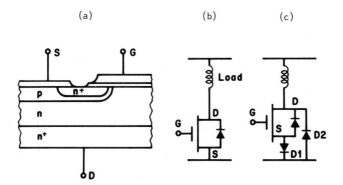

Fig. 10   VMOS antiparallel diode
    (a)   Basic VMOS structure showing built-in diode
    (b)   Circuit with inductive load
    (c)   Solution to charge storage problem by using two external diodes D_1 and D_2.

Measurements on 400 V, 5 A, commercial devices gave Q_s values in the range 1.4 to 5 μC, and t_{rr} of 0.32 to 0.62 μs for the conditions 5 A, 50 A/μs.

When a VMOS device is used as a fast switch in series with an inductive load, an antiparallel diode may be required to recover the energy stored in the inductance. The slowness of the built-in diode will give rise to large currents and power dissipation as the VMOS device switches on. This may destroy the device or, at best, limit the frequency or overall current capability. Ideally, an external diode is needed which has much lower V_F and stored charge. This does not seem possible since VMOS devices already use epitaxial silicon and, in the absence of killing, have a low V_F. One solution is to use two external diodes, as shown.

2. SCHOTTKY DIODES

2.1 Introduction

Schottky diodes are rectifiers which rely for their operation on the barrier formed between a metal in intimate contact with a semiconductor. The barrier to current flow arises from the difference in work functions for the two materials. For power rectification the devices are made from epitaxial n on n$^+$ silicon, with barrier heights, ϕ_B of 0.5 to 0.9 electron volts, depending on the metal chosen. The advantages are two-fold, compared with the normal p-n junction device. Firstly, the low barrier height results in a low voltage in the conducting state. Secondly, the current flow involves only majority carriers, electrons in this instance, and hence there is no stored charge to limit the switching speed. However, the absence of minority carrier injection and conductivity modulation gives rise to a series resistance proportional to the epitaxial layer thickness. In practice this limits the application to reverse voltages of < 50 V. The advantages mentioned make the devices ideal for SMPS with output voltages of < 12 V d.c. and currents of 5 - 100 A.

2.2 Basic Device Concepts

See Ref. 11 for a comprehensive treatment of device physics and further references. Operation of the device can be understood from the band diagram (Fig. 11). With no bias applied the barrier height is sufficient to ensure that the currents of electrons, from the metal to the silicon, and from the conduction band in the silicon back again to the metal, just balance. If a negative voltage V is applied to the silicon the barrier height for electrons flowing to the metal is lowered. This is similar to current flow in a valve, by thermionic emission, and is described by the equation

$$I_F = AR\ T^2\ \exp\ (-\ \frac{q\phi_B}{kT})\ \{\exp\ \frac{qV_F{'}}{kT}\ -\ 1\}$$

where
- A = area of barrier
- B = Richardson's constant = 120 A cm^{-2} K^{-2}
- I_F = forward current
- $V_F{'}$ = forward voltage

This equation can also be written as

$$V_F{'} = \frac{kT}{q}\ \log_e\ \frac{I_F}{I_s},\ \text{for } V_F{'} > \frac{3kT}{q}$$

as for the p-n diode. I_s is much higher however, so that $V_F{'}$ is lower, for a given forward current. An additional component of the forward voltage is due to the series resistance of the epitaxial layer, so that the total V_F is

$$V_F = \frac{kT}{q}\ \log_e\ \frac{I_F}{I_s}\ +\ \frac{I_F\rho t}{A}$$

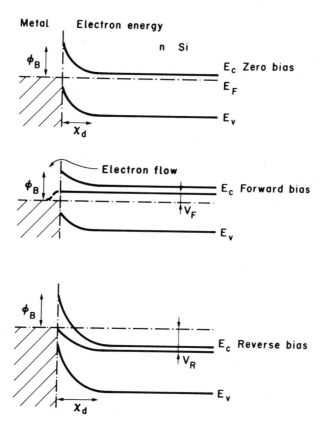

Fig. 11   Schottky diode band diagrams.

where ρ = epitaxial layer resistivity
          t = epitaxial layer thickness

In reverse bias the bands are bent so as to increase the barrier to electron flow from the silicon. The thermionic emission equation still applies, and substituting $V_R > -3kT/q$ for V_F' gives

$$I_R = ART^2 \exp\left(-\frac{q\phi_B}{kT}\right) = I_s$$

i.e. a constant or saturated current. This constant current is due to electrons from the metal which can surmount the barrier and flow to the silicon. It is constant because ϕ_B is constant.

However, the barrier height to electron flow is affected by the presence of electrons in the silicon in the vicinity of the interface. The electrons are attracted to the metal by the image force effect, and hence they have a potential energy, which falls off rapidly into the silicon. This potential modifies the potential barrier, reducing it by an amount $\Delta\phi_B$.

This is dependent on the field at the interface E_s

$$\Delta\phi_B = \sqrt{\frac{q\,E_s}{4\pi\varepsilon_s\varepsilon_o}}$$

where E_s is given in the normal way from the depletion approximation, i.e.

$$E_s = \sqrt{\frac{2qN_D\,(V_R + \phi_B)}{\varepsilon_s\varepsilon_o}}$$

N_D = epi layer doping density
V_R = reverse bias

This lowering of the barrier height increases the reverse leakage as V_R increases.

The breakdown voltage of the structure can be determined in the same way as for the p-n junction epi device, since the Schottky behaves like a one-sided step junction. Alternatively, a one-dimensional calculation of the avalanche integral yields a similar graph to Fig. 3, with a resistivity of 0.9 Ω-cm corresponding to a breakdown voltage of 84 V.

The device characteristics are heavily dependent on the barrier height by the effect on I_s

$$I_s = ART^2 \exp\left(-\frac{q\phi_B}{kT}\right)$$

Thus a low barrier height gives a high leakage current in reverse, and a low forward voltage drop. Conversely, a high barrier device has low leakage but a high V_F. The choice of barrier height is crucial to the successful application of the device. Fig. 12 compares the forward and reverse characteristics of Schottky diodes with various barrier heights against the previously mentioned epi and double-diffused junction diode technologies. The Schottky curves have been calculated using the theory given, and the device particulars shown in the figure.

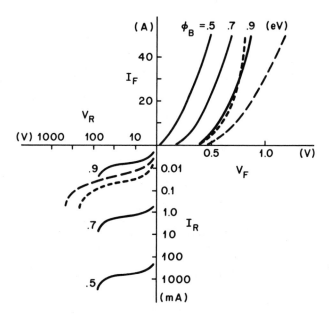

Fig. 12   Comparison of:
      — —400 volt double-diffused diode } (measured)
      — — — —200 volt epi diode
      ————Schottky characteristics (theoretical)
      (Active area = 0.123 cm², Schottky epi: 5 µm 0.9 Ω-cm, T_j = 100°C)

The reverse leakage increases rapidly with temperature. Eventually thermal runaway occurs due to the reverse power dissipation. The lower the barrier height, the lower the temperature at which this occurs. Thus, devices using metals like chromium ($\phi_B \sim$ 0.57) have lower operating temperatures and forward voltages than devices using metals like platinum ($\phi_B \sim 0.87$).

TABLE 2

Schottky Diode Range

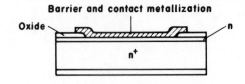

Barrier and contact metallization

Chip details

	$I_{F(AV)}$	Max V_F at $I_{F(AV)}$	Max I_R
	9	0.60	12
	2 x 10	0.60	12
	15	0.60	12
	28	0.55	36
	50	0.55	60
	70	0.55	84
Units	A	V	mA
Notes		1	1

General characteristics

V_{RSM} = 36, 42, 48, 54 V

T_j max   = 150°C

Molybdenum barrier, $\phi_B \sim 0.67$ eV

Notes: 1) at 100°C

Depending on the barrier height, the devices can have similar or higher reverse leakage, and similar or lower forward voltage than the p-n junction devices.

2.3 Typical Properties

Molybdenum has a barrier height of about 0.67 eV, giving characteristics which are optimum for most applications. Table 2 lists these properties for the Philips' range. Other manufacturers make similar devices using nickel platinum (0.70), nichrome (0.69) and tungsten (0.68), as well as molybdenum. The forward and reverse characteristics are predicted accurately by the theory outlined earlier. The Schottky devices are thus easier to model than p-n junctions.

2.4 Suitability for Application

The main application for these devices is switched mode power supplies, where the low forward voltage and absence of stored charge leads to efficient operation at frequencies of 20 kHz or more. However, the lower reverse voltages obtainable with the Schottky technology, i.e. about 50 V, limit the output voltages in practice to about 12 V.

As an example, consider the forward converter of Fig. 5, but with an output current of 50 A at 10 V. When the mains input voltage is high the control circuit will adjust the duty cycle to keep V_o constant at 10 V. Thus, the duty cycle might fall to 0.3 under these conditions. Similarly, with a low mains voltage the duty cycle will rise to the maximum value of 0.45 for this type of circuit. This range of duty cycle will allow for mains input voltage variations of $\pm$ 20 %.

Consider the case for $\delta = 0.45$. The equations in Fig. 5 give the D2 current and voltage waveforms in Fig. 13. The designer of such a power supply may ask the question "What barrier height gives the best performance for D2, considering efficiency, thermal stability margin, device operating temperature and the size heat sink required?"

The equations in B2 allow us to calculate these factors. Let us assume that a D04 diode of active area 0.1225 cm^2 is used. The epi layer, for a practical 50-volt device, might be 0.9 Ω-cm, 5 μm, on a substrate of 2 mΩ-cm, 250 μm thick. These resistivities are themselves functions of temperature, the layer resistivity varying as $\sim T^{2.3}$, due to the fall in mobility as T rises. The thermal resistance from junction to heat sink is about 1.3°C/W for such a device, and then from heat sink to ambient values of 1 to 10°C/W can be considered, depending on the size of the heat sink. Finally, an ambient temperature of 40°C might be found in the power supply.

The power in the device is given by the sum of the forward and reverse dissipations,

$$P = I \cdot V_F \cdot \delta + I_R \cdot V_R \ (1 - \delta)$$

The junction temperature is given by

$$T_j = P \ (1.3 + H) + 40°C$$

Where H is the thermal resistance of the heat sink, and the heat-sink temperature, where the device is mounted by

$$T_{hs} = HP + 40°C$$

Using these equations and solving numerically for values of barrier height ϕ_B from 0.5 to 0.9, and H from 1 to 10, gives the graph of T_j as a function of H, shown in Fig. 13.

It can be seen that for high barrier heights:
a) Operating temperatures can be high
b) Thermal stability is good
c) Power loss is high

A barrier height of 0.7 is a good compromise between the conflicting requirements, as can be seen from the table for a heat-sink size of 3°C/W. A similar conclusion can be drawn from a value of $\delta = 0.3$. In general high reverse voltages and low currents favor higher barrier devices, since thermal runaway is caused by reverse dissipation, while low voltages and high currents favor low barrier diodes, because of the low forward losses.

The epi diode is a better choice, in this application, than devices with $\phi_B > 0.8$ eV, and would, of course, give much higher reverse voltage protection. The only limitation to higher frequency operation is the device capacitance. This is due to the depletion layer, and has the usual form

$$C = \sqrt{\frac{q\varepsilon_s\varepsilon_o \ N_D}{2(V_R + \phi_B)}}$$

where C = small signal capacitance per unit area
   N_D = epi doping level
   V_R = reverse voltage

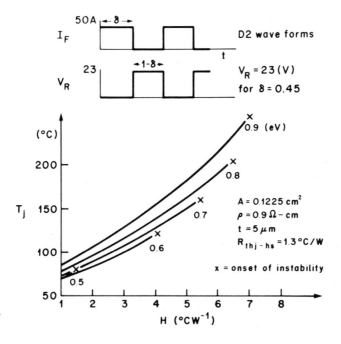

For H = 3 °C W^{-1}

Barrier height	0.5	0.6	0.7	0.8	0.9	Epi diode
Power loss W	-	14.0	15.9	18.3	20.9	18.4
T_j °C	-	100	108	119	130	119
Stability margin	-	25	54	90	120	56

Fig. 13   50 A, 10 V SMPS using Schottky diodes. Performance of D2 as a function of barrier height.

This capacitance is thus difficult to lower, since lower values of N_D will badly affect the series resistance of the device. However, in practice, this is only likely to be a problem for frequencies of more than 100 kHz.

Two-dimension modelling of the edge structure used has recently been carried out with the program SEMMY.[12],[13] This shows that it is possible to get a high breakdown voltage at the edge of the device using a simple field plate structure. The two-field peaks, at the end of the field plate and at the edge of the oxide window, have to be balanced in order to optimize the structure. Fig. 14 shows a plot of the equipotentials for a device with an oxide layer which is too thick. The crowding of the potential lines at the edge of the oxide window gives rise to a high field, lowering the breakdown voltage. Plots of the field distribution and calculation of the avalanche integral are also available from this program, and in this case indicated that the device could be optimized using a thinner oxide layer.

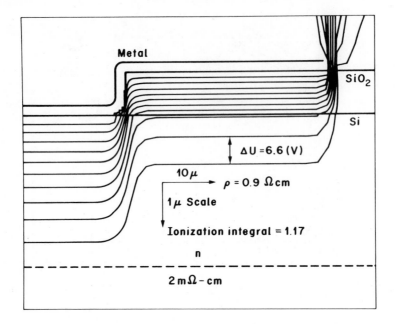

Fig. 14   Calculated equipotentials within a device with too thick an oxide layer.

2.5 Goals for Future Development

The available devices have characteristics which are "ideal", that is, they correspond closely to the theory. The technology is maturing, and large yield-related, cost reductions are not expected. However, the integration of two or more diodes, and the use of plastic packages will lead to cheaper devices.

The efficiency of the devices in the application is largely dependent on correct choice of barrier height, as pointed out earlier. For example, the introduction of logic running from 2-volt supplies may require low barrier devices (e.g. chromium).

3. EXTENSION OF EPI AND SCHOTTKY DIODE
TECHNOLOGY TO HIGHER POWERS

There seems to be no fundamental reason why the device technologies cannot be extended to higher currents, certainly by an order of magnitude to ∿ 1000 A. Of course the normal high-power packaging methods would have to be used, e.g. pressure contacts. The quality of epitaxial material currently available is excellent, posing little obstacle to devices of ∿ 10 cm^2. Such devices, particularly Schottky diodes, may find application in rectification at 50 Hz, where the low forward voltage would increase efficiency. Losses would remain in the 50-Hz transformer.

However, there are difficulties in the application of high-frequency switching techniques above about 100 A. The design of the switching devices and the transformers presents problems. The largest practical switched mode power supply that can be considered at the moment is about 10 kW, e.g. 24 V at 400 A. This could use two 200-A epi diodes. However, in practice four 100-A diodes would be equally easy to use, and might simplify the heat sinking.

Another question is the purpose of such high current, low voltage supplies. Welding is one possibility. Some requirements, such as very large computers, can more easily use many local smaller supplies, to avoid the difficulties in distributing high currents around the equipment.

REFERENCES

1. J.A.F. Cornick, Internal Memorandum (January 1979).

2. A. Herlet, S.S.E. 11 (1968) 717.

3. H. Schlangenotto et al., Forschungsbericht Bundesministerium für Forschung und Technologie BMFT-FB-T 76-54-Bonn (1975).

4. F. Berz, S.S.E. 20 (1977) 709.

5. H. Schlangenotto and H. Maeder, "Spatial Composition and Injection of Recombination in Silicon Power Device Structures," IEEE Trans. Electron Devices, ED-26 (March 1979) 191-200.

6. J.A.F. Cornick, Internal Memorandum (March 1981).

7. Mullard Technical Information No. 41, Mullard House, Torrington Place, London WC1E 7HD.

8.   M.D. Miller, "Differences between Platinum- and Gold-Doped Silicon Power De-
     vices," IEEE Trans. Electron Devices, ED-23 (December 1976) 1279-1283.

9.   S.D. Brotherton and P. Bradley, S.S.E. Paper in press.

10. M. Derdouri et al., "A Comparative Study of Methods of Measuring Carrier Life-
     time in p-i-n Devices," IEEE Trans. Electron Devices, ED-27, Part. II (November
     1980) 2097-2101.

11. E.H. Rhoderick, Metal Semiconductor Contacts, Oxford: Clarendon Press, 1978.

12. Article to be published in International Journal of Numerical Methods in Engineer-
     ing (Wiley).

13. NASECODE, Dublin, 1979.

DISCUSSION

(Chairman: Dr. H. Becke, Bell Laboratories, Murray Hill, N.J., USA)

DISCUSSION
(Chairman: Dr. H. Becke, Bell Laboratories, Murray Hill)

H. Becke

Do you use a special structure like a p-n junction or a ring around the Schottky diodes?

R. Grover (Mullard, Hazel Grove)

We see no real advantage to the use of guard rings and so don't use them. As I showed in the two-dimensional plot you can get quite good reverse voltages with a simple field-plate structure and it's in the nature of the Schottky device to be rather susceptible to any high temperature processing. It's operating right on the surface of the silicon and the fewer high-temperature processes you introduce the better.

H. Becke

Were the surface recombination velocities in the vicinity of the outside edge taken into account in the model of a two-dimensional field distribution?

R. Grover

No, not in that model.

A. Neidig (Brown, Boveri & Cie. AG, Lampertheim)

Is there any temperature treatment after depositing the molybdenum and how do you maintain the barrier height over the larger surfaces when you try to produce diodes of larger diameters?

R. Grover

There are no special annealing treatments for the molybdenum/silicon interface. I can't really answer the second part of your question because the largest diodes we've made are 6 mm square and we don't see any variation in barrier height for those small devices. I agree it could be a problem in making larger devices.

M. Adler (General Electric, Schenectady)

You showed a figure of the waveforms of both diodes in a circuit which demonstrated that the Schottky is ringing more than the epi diode. Does the capacitance of a Schottky diode limit its application at high frequencies and how can it be compared to the capacitance of p-n junction devices?

R. Grover

As Schottkys go to higher and higher frequencies this capacitance will become a difficulty and may be a barrier for switching frequencies of 100 kHz or more.

M. Adler

Is the capacitance of Schottky diodes higher because they're definitely bigger devices for the same current?

R. Grover

The Schottky device has to use the absolute minimum resistivity compatible with the reverse voltage because it's a majority carrier device, whereas in the epitaxial device you've got the freedom to use higher resistivity silicon and this leads to a reduction in capacitance for equivalent devices.

A. Jaecklin (BBC Brown, Boveri & Co. Ltd., Baden)

It has been frequently reported in the literature that you can enhance the reverse-blocking capability of Schottky diodes with tapered silicon oxide layers. What is your opinion on that?

R. Grover

In the slide there was a slight degree of tapering on the edge of the oxide window. I think it's beneficial.

A. Jaecklin

What would be the limits in voltage?

R. Grover

I find it difficult to give you a figure for that. That semi-calculation showed the device breaking down at 70 V. The one-dimensional calculation of the avalanche integral says that the structure should break down at round about 85 V, which is not too far away.

OPEN DISCUSSION

(Chairman: Dr. H. Becke, Bell Laboratories, Murray Hill)

H. Becke

In this discussion period we should like to see whether we can agree as to the purposes for which a given device is best suited and to compare GTOs with bipolar
transistors, FETs, static-induction thyristors and transistors.

M. Adler (General Electric, Schenectady)

I would like to ask Dr. Tihanyi whether he feels that the MOS-controlled bipolar
transistor that he showed will find an application or whether in fact one would be
better off with a normal bipolar Darlington in terms of area utilization?

J. Tihanyi (Siemens AG, Munich)

These devices especially are better than the bipolar Darlingtons because they exhibit a truly MOS input. The breakdown voltage, however, is a bipolar emitter to
collector-breakdown and does not correspond to the full blocking capability of the epitaxial layer. Perhaps the ideal solution would be to use a MOSFET connected in parallel with a bipolar transistor. The MOSFET is switched on first at high voltage,
and then the bipolar transistor is turned on offering a lower resistance in the on-
state. For turn-off the bipolar has to be switched off first. The MOSFET leads the
current again for a period long enough to sufficiently reduce the carrier concentration in the bipolar transistor and then the MOSFET is turned off. Up to now nobody
has achieved such a solution.

H. Becke

Do you mean that there is a connection in the input to the transistor base as well as
to the MOS-gate in such a manner that the devices are not strictly in parallel but
that the MOS is driven with a certain delay when the storage time of the transistor
is over? Or are they just strictly in parallel?

J. Tihanyi

I wouldn't say that this solution is actually feasible, but it represents the direction
in which we think at least. IC technology offers several possibilities for achieving
such an integration.

H. Becke

Do you expect this technique to be applied to a large-area devices?

J. Tihanyi

Yes, however, in comparison to high-power thyristors the MOS transistors are too fine. We have to make the whole technology somewhat coarser.

H. Becke

There is definitely a relationship between optimum cell size and blocking voltage. For a 1000-V power MOS, it makes no more sense to make a 20µm-cell when the depletion-layer width will be about 2 or 3 times as wide. For devices of 30 or 40 mm side lengths it is just a question of whether you can achieve 100% yield, because you must not have a single short cell.

J. Tihanyi

That is the reason why the GTOs using the finer geometry like SITs are more sensitive than the other ones.

H. Becke

Dr. Nishizawa, you use 10µm spacing, which is a very fine structure. Wouldn't you anticipate problems for large areas?

J. Nishizawa (Tohoku University, Sendai)

Nowadays the processing technology has reached a very high standard. In our case we need an accuracy of a few microns, which represents an easy task for VLSI people. In fact, these devices can be and are prepared even by students.

D. Silber (AEG-Telefunken, Frankfurt)

I would like to come back to the Darlington consisting of a MOSFET and a bipolar transistor. I think a speed-up diode is required to obtain high-switching speed and so triggering becomes more complicated.

M. Adler

The thing that concerns me about that structure is that any voltage drop across the MOS transistor adds directly to the output voltage of the bipolar transistor. To achieve similar output drops to a bipolar Darlington with that structure you find that the MOS transistor is as large as the output stage. Furthermore dealing with basic-ally a bipolar device as the output device you are going to be limited by its speed. So you're paying dearly for the MOS first stage in that device. I'm skeptical that that structure is really viable in many applications. The simple Darlington connection offers the simplicity of the MOS gate, but wastes the speed of the front-end device. In a sense the parallel combination makes use of both devices and that would be a little different.

R. Sittig (BBC Brown, Boveri & Co. Ltd., Baden)

Prof. Nishizawa stated that tody's technology permits the production of complicated junction field-effect devices. I agree, but this is only one side of the coin. In general these devices will not open up new fields of application but will eventually penetrate into the same applications as present power devices. This means, that in the development of power devices we are not fighting against physical limitations in the first place but against costs. The question, therefore, is whether an installation using field-effect devices is lower in total cost compared to the present solution, and this in turn depends on characteristics like production yield and reliability during operation.

J. Tihanyi

That's all true if you don't take into account the higher speed. Higher frequencies offer really new possibilities for power-electronics applications.

M. Adler

I agree completely as far as increased frequency performance for the MOSFET is concerned. But the MOS Darlington doesn't give you any improvement in frequency capability.

R. Sittig

We should agree on the meaning of "frequency". Most important in power electronics is the switching frequency, which can be applied in long-term operation. Using this definition and considering an operation at nearly breakdown voltage switching losses will very soon limit the frequency with present thermal resistances. As I indicated in my paper, generally even the losses occurring in the source-to-drain capacitance of a 400-V device will be too high to operate it at 1 MHz. Therefore I expect an extension of frequency only in the 10 to 100 kHz range for devices of about 1 kV blocking voltage.

H. Becke

One can try to sketch a diagram with frequency as the x-axis and the volt-amp product cost as the y-axis. The MOSFETs, SITs, bipolar transistors, GTOs and thyristors will appear in some sequence from highest frequency and relative cost to the lowest values. One has, however, to consider other influences on this cost-to-performance chart. One is the temperature. The maximum possible temperature for transistors is 150-175°C and for thyristors it is about 125-130°C, while the characteristics of MOSFETs are mainly calculated for 80°C and degrade considerably for higher temperatures. Furthermore there may be other factors like the reverse second-breakdown phenomenon, which favors the MOSFET compared to transistors. It is always necessary to consider the system as a whole and then see where there is a tradeoff and what optimum can be achieved.

F. Schwarz (Delft University of Technology, Delft)

I agree with Dr. Sittig that the main objective is to bring the cost down. A very important aspect of equipment cost is the passive components. The size of the passive components depends to a large extent on the internal frequency of operation. An increase in the capability of operating at higher frequencies is one key to lowering the cost of the equipment being produced. In addition to the search for new and better mechanisms in semiconductors, this point should represent a significant part of the quest to improve the semiconductor technology.

M. Depenbrock (Ruhr University, Bochum)
I would like to ask Dr. Becke why he believes that the GTO has better high-frequency properties than the normal thyristor since the frequency capability depends mainly on the dI/dt, dV/dt and the switch-on properties but not on the turn-off ability.

H. Becke
That's correct, but when you talk about GTO then immediately you talk about negative bias on the gate which can be held during the off state, and this in effect gives you a dV/dt which is several thousand volts per microsecond. You could do this with a thyristor, but you cannot do it with a thyristor, which has a shorted emitter.

P. de Bruyne (BBC Brown, Boveri & Co. Ltd., Baden)
Dr. Becke, you gave the impression that GTOs can be operated up to 200 kHz. Is this realistic, and for which power level is it possible?

H. Becke
I assume this can be extended to the sizes to which normal thyristors will go. Certainly for currents up to 50 A, I do not see a problem to build a 50- or 100-kHz inverter, for example.

P. de Bruyne
I imagine you are mainly speaking about GTOs of relative low power, for example up to 1000 V and 10 A. For these devices a cathode area occupying only 30% of the total area is easily tolerated. But if we consider the situation for a high-voltage thyristor for 600 A, which is presently produced from a 2-inch wafer, for a corresponding GTO I would expect an increase in cost by at least a factor of three.

H. Becke
Although I share your opinion, this may not be quite right. If you want to have a thyristor which goes up into the 100-kHz switching range, then you must use an interdigitated gate structure and the situation for a regular thyristor becomes similar to that of a GTO. Since the anode area of the GTO is larger than the cathode, you get current and heat spreading in the vertical direction. This allows the GTO to be operated at a higher current density referred to the cathode area. Therefore the re-

lative loss of active area will be considerably smaller than the 70% which you mentioned.

P. de Bruyne

From the values which Prof. Nishizawa reported in his paper for the static induction transistor, I have calculated an average current density of 15 A/cm^2 referred to the total area. This is an order of magnitude less than the values which you mentioned for the GTOs. In addition, I am not sure whether the vertical heat spread will result in a reduction of the thermal resistance. Since in a GTO with a press-pack case only about 30% of the cathode-side surface is cooled, I would expect an increase in the thermal resistance. Perhaps Dr. Kurata could give the actual values which he observed for high-power GTOs.

M. Kurata (Toshiba R & D Center, Kawasaki)

We have concerned ourselves mostly with low-frequency high-power GTOs, which can be operated at about 1 kHz. To that extent we can work with GTO's and this takes into account the thermal resistance of the package. I cannot, however, give the limits of frequency capability of GTOs. I do feel that Dr. Becke may be too optimistic.

H. Becke

This we shall find out in the future. I can say, however, that I have operated devices with 200 A/cm^2 referred to the total cross section of the device, and that there are methods, such as dynamic balasting which allow the current distribution to be kept uniform in such a way that there is no limitation in total area.